Marine Oil Spills

Special Issue Editor
Merv Fingas

MDPI • Basel • Beijing • Wuhan • Barcelona • Belgrade

MDPI

Special Issue Editor
Merv Fingas
Spill Science
Canada

Editorial Office
MDPI AG
St. Alban-Anlage 66
Basel, Switzerland

This edition is a reprint of the Special Issue published online in the open access journal *Journal of Marine Science and Engineering* (ISSN 2077-1312) from 2015–2016 (available at: http://www.mdpi.com/journal/jmse/special_issues/marine_oil_spills).

For citation purposes, cite each article independently as indicated on the article page online and as indicated below:

Author 1; Author 2. Article title. *Journal Name* **Year**, *Article number*, page range.

First Edition 2017

ISBN 978-3-03842-504-5 (Pbk)
ISBN 978-3-03842-505-2 (PDF)

Table of Contents

About the Special Issue Editor v

Preface to "Marine Oil Spills" vii

Krishna K. Kadali, Esmaeil Shahsavari, Keryn L. Simons, Petra J. Sheppard and
Andrew S. Ball
RNA-TGGE, a Tool for Assessing the Potential for Bioremediation in Impacted Marine
Ecosystems
Reprinted from: *J. Mar. Sci. Eng.* **2015**, 3(3), 968–980; doi: 10.3390/jmse3030968....................................1

Mace G. Barron, Jill Awkerman and Sandy Raimondo
Oil Characterization and Distribution in Florida Estuary Sediments Following the
Deepwater Horizon Spill
Reprinted from: *J. Mar. Sci. Eng.* **2015**, 3(3), 1136–1148; doi: 10.3390/jmse3031136...........................12

Tonya Cross Hansel, Howard J. Osofsky, Joy D. Osofsky and Anthony Speier
Longer-Term Mental and Behavioral Health Effects of the Deepwater Horizon Gulf Oil Spill
Reprinted from: *J. Mar. Sci. Eng.* **2015**, 3(4), 1260–1271; doi: 10.3390/jmse3041260...........................23

Nusrat Jahan, Jason Fawcett, Thomas L. King, Alexander M. McPherson, Katherine
N. Robertson, Ulrike Werner-Zwanziger and Jason A. C. Clyburne
Bitumen on Water: Charred Hay as a PFD (Petroleum Flotation Device)
Reprinted from: *J. Mar. Sci. Eng.* **2015**, 3(4), 1244–1259; doi: 10.3390/jmse3041244...........................32

Gloria Biern, Jesús Giraldo, Jan-Paul Zock, Gemma Monyarch, Ana Espinosa,
Gema Rodríguez-Trigo, Federico Gómez, Francisco Pozo-Rodríguez, Joan-Albert Barberà
and Carme Fuster
Human Genotoxic Study Carried Out Two Years after Oil Exposure during the Clean-up
Activities Using Two Different Biomarkers
Reprinted from: *J. Mar. Sci. Eng.* **2015**, 3(4), 1344–1348; doi: 10.3390/jmse3041334...........................45

Juan M. Restrepo, Jorge M. Ramírez and Shankar Venkataramani
An Oil Fate Model for Shallow-Waters
Reprinted from: *J. Mar. Sci. Eng.* **2015**, 3(4), 1504–1543; doi: 10.3390/jmse3041504...........................57

Ann Hayward Walker, Clay Stern, Debra Scholz, Eric Nielsen, Frank Csulak and
Rich Gaudiosi
Consensus Ecological Risk Assessment of Potential Transportation-related Bakken and
Dilbit Crude Oil Spills in the Delaware Bay Watershed, USA
Reprinted from: *J. Mar. Sci. Eng.* **2016**, 4(1), 23; doi: 10.3390/jmse4010023...........................91

Susan Laramore, William Krebs and Amber Garr
Effects of Exposure of Pink Shrimp, *Farfantepenaeus duorarum*, Larvae to Macondo Canyon 252
Crude Oil and the Corexit Dispersant
Reprinted from: *J. Mar. Sci. Eng.* **2016**, 4(1), 24; doi: 10.3390/jmse4010024...........................117

Kristen Shapiro, Shruti Khanna and Susan L. Ustin
Vegetation Impact and Recovery from Oil-Induced Stress on Three Ecologically Distinct
Wetland Sites in the Gulf of Mexico
Reprinted from: *J. Mar. Sci. Eng.* **2016**, *4*(2), 33; doi: 10.3390/jmse4020033..135

Lizabeth Bowen, A. Keith Miles, Brenda Ballachey, Shannon Waters and James Bodkin
Gene Transcript Profiling in Sea Otters Post-Exxon Valdez Oil Spill: A Tool for Marine Ecosystem
Health Assessment
Reprinted from: *J. Mar. Sci. Eng.* **2016**, *4*(2), 39; doi: 10.3390/jmse4020039..154

About the Special Issue Editor

Merv Fingas is a scientist working on oil and chemical spills. He was Chief of the Emergencies Science Division of Environment Canada for over 30 years and is currently working on research in Western Canada. Dr. Fingas has a PhD in environmental physics from McGill University, three Masters degrees— in chemistry, business and mathematics—all from the University of Ottawa. He also has a bachelor of science in Chemistry from Alberta and a bachelor of arts from Indiana. He has published more than 900 papers and publications in the field. Merv has prepared nine books on spill topics and is working on another. He is chairman of several ASTM and inter-governmental committees on spill matters. Dr. Fingas's specialities include oil chemistry, spill dynamics and behaviour, spill treating agents, remote sensing and detection, and in-situ burning. He continues to work on developing new technologies in these fields.

Preface to "Marine Oil Spills"

Oil spill studies continue to evolve. While there are few books on the topic, there are regular conferences and symposia. This is one of the few books on the topic of oil spills. As such, this book focuses on providing material that is quite diverse and covers a wide variety of topics. Most of the content of this book focuses on studies that were initiated following the Deepwater Horizon spill. This spill was devastating in many ways. The Deepwater spill was the largest spill in the USA. The cost was immense both in terms of dollars and in social terms. The spill caused many academics to focus on the problem of oil spills using their own expertise and tools. The ten studies presented in this book are examples of such studies and present a cross-section of the variety of the hundreds of studies that followed the Deepwater Horizon spill. The studies following the Deepwater Horizon spill have opened up many new fields of spill studies and in turn much knowledge about the fate, effects and control of oil spills.

Merv Fingas
Special Issue Editor

Journal of
*Marine Science
and Engineering*

MDPI

Article

RNA-TGGE, a Tool for Assessing the Potential for Bioremediation in Impacted Marine Ecosystems

Krishna K. Kadali [1,2], Esmaeil Shahsavari [1,2], Keryn L. Simons [2], Petra J. Sheppard [1,2] and Andrew S. Ball [1,2,*]

[1] Centre for Environmental Sustainability and Remediation, School of Applied Sciences, RMIT University, PO Box 71, Bundoora 3083, Australia; kadalikishore@gmail.com (K.K.K.); esmaeil.shahsavari@rmit.edu.au (E.S.); petra.reeve@sawater.com.au (P.J.S.)
[2] School of Biological Sciences, Flinders University, GPO Box 2100, Adelaide 5001, Australia; keryn.simons@flinders.edu.au
* Author to whom correspondence should be addressed; andy.ball@rmit.edu.au; Tel.: +61-03-99256594; Fax: +61-03-9594-7110.

Academic Editor: Merv Fingas
Received: 18 June 2015; Accepted: 26 August 2015; Published: 31 August 2015

Abstract: Cultivation-independent genomic approaches have greatly advanced our understanding of the ecology and diversity of microbial communities involved in biodegradation processes. However, much still needs to be resolved in terms of the structure, composition and dynamics of the microbial community in impacted ecosystems. Here we report on the RNA activity of the microbial community during the bioremediation process using RNA Temperature Gradient Gel Electrophoresis (RNA-TGGE). Dendrograms constructed from similarity matching data produced from the TGGE profiles separated a community exhibiting high remediation potential. Overall, increased Shannon Weaver Diversity indices (1–2.4) were observed in the high potential remediation treatment samples. The functionality of the microbial community was compared, with the microbial community showing the greatest organisation also showing the highest levels of hydrocarbon degradation. Subsequent sequencing of excised bands from the microbial community identified the presence of *Gammaproteobacteria* together with a number of uncultured bacteria. The data shows that RNA TGGE represents a simple, reproducible and effective tool for use in the assessment of a commercial bioremediation event, in terms of monitoring either the natural or augmented hydrocarbon-degrading microbial community.

Keywords: bioremediation; microbial community dynamics; Pareto-Lorenz curve; RNA; Shannon Weaver diversity; TGGE

1. Introduction

In the past, detection and analysis of bacteria in the environment was performed mainly by methods based on bacterial culture [1]. As widely reported, the application of this technique leads to the isolation only of around 0.001% of the microbial population present in sea water and 0.3% present in soil [2,3]. In terms of commercial bioremediation and the management of a bioremediation event, this technology has only been of limited value due to the length of time required for isolation [4]; on many occasions by the time a drop in hydrocarbonoclastic organisms has been observed [5], the bioremediation has already stalled.

Developments in cultivation-independent genomic approaches have greatly advanced our understanding of the ecology and diversity of microbial communities [6]. For example, the separation or detection of small differences in specific DNA sequences can give important information about community structure and the diversity of microbes containing critical genes [4,7]. Many fingerprinting techniques have been developed and used in applied microbial ecology situations, such as bioremediation [1,4,8]. Molecular genetic fingerprinting techniques provide a pattern or profile of the community diversity on the basis of the physical separation of unique nucleic acid species. The general strategy for genetic fingerprinting of microbial communities consist of first, the extraction of nucleic acids (DNA and RNA), second the amplification of genes encoding 16S rRNA and third, the analysis of PCR products by a genetic fingerprinting technique [9,10]. Different amplified products can be separated by electrophoresis to create banding patterns known as a molecular fingerprint. Changes in the molecular fingerprint can be analysed to identify the microbial community structure in space and in time [11].

TGGE (Temperature Gradient Gel Electrophoresis) is an established community profiling tool that allows the study of the complexity and behaviour of microbial communities [9]. TGGE separates the PCR amplified DNA fragments (200–700 bp) of the same length but with different sequences [8] like DGGE (Denaturing Gradient Gel Electrophoresis). TGGE uses temperature gradient to separate DNA fragments. In TGGE, the use of a controlled temperature gradient simplifies the experiment and leads to reproducible gel results [12].

Genomic DNA (16S rDNA) has been widely used in fingerprinting techniques like TGGE to study the microbial community in various environments including petroleum contaminated soils [13], waste water treatment [14], streams and rivers [15] and marine waters [16]. DNA based TGGE is popular because the extraction protocols of DNA are simple and DNA samples can be readily handled. However, genomic DNA may not always be considered a suitable technique because detection of the DNA neither indicates the activity nor proves the viability of cells [17,18]; DNA can persist for long periods of time in the environment after the cells have lost viability [19]. The alternative approach is to use is RNA. rRNA sequences have been used as a marker for bacterial activity since the amount of ribosomes (and their rRNA) per cell was found to be roughly proportional to the growth and activity of bacteria in pure cultures [20]. It was also reported that extraction of RNA instead of DNA followed by reverse transcription polymerase chain reaction gives information on the metabolically active microbial community [4]. Therefore, fingerprints based on RNA better represent the most abundant as well as the more active populations [21]. However, limitations to successful RNA based TGGE exists, including difficulty in extracting intact RNA [22,23]. To date there is limited research in terms of RNA based TGGE [18,24]. The role of many bacteria in the natural environment remains unknown and our knowledge about the structure, composition and dynamics of the microbial community that inhabits impacted ecosystems is still lacking [25]. Recent progress in metagenomic approaches such as next generation sequencing (NGS) may lead to a better understanding of microbial communities involved in hydrocarbon degradation in marine environments. However, these methods are still labour intensive, and expensive. For example, metagenomics generates huge amounts of data which needs high-performance computing and automated software whereas RNA-TGGE fingerprinting as a simple routine technique may be more suitable for monitoring a bioremediation project.

The aim of this study was to analyse the 16S rRNA amplicons based on TGGE from mesocosms which examined the role of bioaugmentation and/or biostimulation on the extent of mineralisation of weathered crude oil in sea water and study their diversity and functionality through Parento-Lorenz curves. We suggest that this is a potentially important tool for use in the assessment of a bioremediation event, in terms of monitoring either the natural or augmented hydrocarbon-degrading microbial community. To date, certainly in the remediation industry, the application of molecular microbial ecological techniques has not been widely adopted. This study highlights the potential of RNA-TGGE as a reproducible tool for the monitoring of assessing biodegradation potential during bioremediation.

2. Materials and Methods

2.1. Sample Collection

Sea water and weathered crude oil samples were collected from the three treatments C (seawater + BH medium + weathered crude oil + consortia), O (seawater + BH medium + weathered crude oil) and S (seawater + BH medium) obtained from previous work [26] in which consortia comprised of six bacterial strains grown separately and mixed (final OD$_{600}$ 0.04). Samples were collected from the oil-medium interface using sterile 10 mL tubes. All the samples used were in duplicates. Samples were further analyzed sequentially using molecular techniques as described below. All the sea water used for the experiment was freshly collected from the South Australian coast.

2.2. Nucleic Acid Extraction

Glass beads (0.5 g) (212–300 µm) were measured into 2 mL tubes and sterilised at 121 °C for 15 min. Samples (600 µL) and stool lysis buffer (800 µL) (Qiagen, Hilden, Germany) were added into the sterile tubes containing glass beads. The tubes were bead beaten for 1 min using a mini-bead beater (Biospec Product, Bartlesville, OK, USA), then incubated for 6 min at 70 °C. After incubation the tubes were centrifuged at 13,000 rpm at 4 °C for 2 min. The supernatant from the tubes was separated into sterile 2 mL Eppendorf tubes (Melbourne, Australia) and equal volumes of phenol chloroform (1:1 ratio) added. Eppendorf tubes were centrifuged at 13,000 rpm at 4 °C for 2 min. The supernatants were separated into new sterile Eppendorf tubes and again equal volumes of phenol-chloroform were added. The tubes were re-spun at 13,000 rpm at 4 °C for 2 min. Supernatants were placed into sterile Eppendorf tubes containing an equal volume of cold iso-propanol. The tubes were incubated at −20 °C for 1 h. After incubation the tubes were centrifuged at 13,000 rpm at 4 °C for 12 min. Supernatants were discarded and an equal volume of cold ethanol added. After mixing the tubes were centrifuged for 13,000 rpm at 4 °C for 12 min. The supernatant was discarded and the pellet was allowed to dry. Finally the pellet was dissolved in nuclease-free water (50 µL) and stored at −20 °C.

2.3. DNase Treatment of RNA Samples

RNA samples were treated with using RQ1 (RNase free DNase, Promega, Melbourne, Australia) according to the manufacturer's guidelines.

2.4. cDNA Synthesis

The first cDNA strand was synthesised using a two-step process (Promega, Melbourne, Australia). In the first step 14 µL reaction mixture containing RNA (8 µL), reverse primer 518R (2 µL) (10 pmol/µL) and sterile nuclease-free water was incubated at 70 °C for 8 min and cooled on ice quickly for 5 min. In the second step the 25 µL reaction mixture containing RNA (14 µL) (from step 1), M-MLV RT buffer (5 µL) (5×), deoxynucleoside triphosphate (dNTP) mixture (1.25 µL) (10 mM), M-MLV reverse transcriptase (1 µL) (100U/µL) and sterile nuclease-free water was incubated at 55 °C for 60 min followed by 70 °C for 15 min to get final cDNA.

2.5. Bacterial cDNA Amplification

The active bacterial community (cDNA) was evaluated by PCR using universal primers of 16S rDNA using the following primers, 341F (5′ CCTACGGGAGGCAGCAG 3′) with GC clamp (CGCCCG CCGCGCGCGGCGGGCGGGGCGGGGGCACGGGGGG) and 518R (50-ATTACCGCGGCTGCTG G) [10]. The PCR amplification of bacterial cDNA was performed in a 50 µL polymerase chain reaction (PCR) mixture. The master mix contained forward primer (2 µL) (10 pmol/µL), reverse primer (2 µL) (10 pmol/µL), magnesium chloride (3 µL) (25 mM), deoxynucleoside triphosphate (dNTP) mixture (1 µL) (10 mM), GoTaq flexi buffer (10 µL) (5×), Taq polymerase enzyme (0.25 µL) (5U/µL) and sterile nuclease-free water per PCR reaction. The cDNA extract (2 µL) was added to 48 µL of master mix. The thermocycling program used consisted of one cycle 5 min at 95 °C; 4 cycles of 30 s at 94 °C, 30 s at 55 °C, 1 min at 72 °C; 25 cycles of 30 s at 92 °C, 30 s at 55 °C, 1 min at 72 °C; and a final extension at 72 °C for 10 min. All the reagents were obtained from Promega, Melbourne, Australia.

2.6. TGGE (Temperature Gradient Gel Electrophoresis)

Products obtained from PCR were analysed by TGGE, using a TGGE Maxi System (Biometra, Germany). Gels were composed of 6% acrylamide (37:5:1). Polymerization of gels were catalysed by the addition of N,N,N',N'-tetramethylethylenediamine (50 µL) and 10% ammonium persulfate solution (500 µL) added to the gel solution (50 mL). The gel was loaded with products amplified with 341F-GC and 518R primers (8 µL) and a dye solution (2 µL) and run for 8 h at 250 V with a parallel temperature gradient ranging from 45 to 60 °C. After electrophoresis, the gel was silver stained [27]. Gel images were scanned with an Epson V700 scanner (Epson, Melbourne, Australia).

2.7. Identification of Bacterial Species from TGGE Gel

Bands of interest on the TGGE gel was excised aseptically with sterile scalpels and incubated in sterile nuclease-free water (100 µL) overnight at 4 °C. The eluted DNA was subjected to PCR using primers 341F and 518R.

PCR reactions were purified using a PCR clean up kit (Promega, Melbourne, Australia) and quantified by Nanodrop (Thermo Fisher Scientific, Melbourne, Australia). The samples were then sent for sequencing to AGRF (Australian Genome Research Facility) according to AGRF requirements. Chromatographs of the sequences received from the AGRF were checked, edited and assembled using Sequencher software (Version 4.9). The aligned sequences were analysed using the nucleotide BLAST program. Species were matched with highest identity scoring.

2.8. Statistical Analyses

Relative band intensities or peaks on TGGE community profiles were calculated using Phoretix 1D advanced analysis package (Totallab, Newcastle, UK). Each band was considered to be an operational taxonomic unit (OTU) and the band densities were then used to calculate the Shannon Weaver diversity and Pareto-Lorenz curve [28]. For each TGGE lane, the respective bands were ranked from high to low based on their intensities; subsequently, the cumulative normalized number of bands was used as X-axis, and their respective cumulative normalized intensities represent the Y-axis.

3. Results and Discussion

3.1. Microbial Community Dynamics

The samples collected were used to extract rRNA and subsequent cDNA were produced using primers 341F-GC and 518Rfor use in the TGGE gel. TGGE gel profiles were used to construct an UPGMA dendrogram (Figure 1) from similarity matching data. The samples were C (seawater + BH medium + weathered crude oil + consortia), O (seawater + BH medium + weathered crude oil) and S (seawater + BH medium). Sample C represented a high remediation capacity community, able to significantly degrade weathered crude oil (28% degradation). Sample O represented a microbial community with a lower remediation potential in terms of degradation of weathered crude oil (16%). Sample S represented the microbial community of a pristine sea water sample which was amended with nutrients but was not contaminated with petrogenic hydrocarbons [26].

The amplified PCR products from Week 0 through to Week 4, of all three different treatment samples (C, O and S) were analysed by TGGE. Generally at Week 0, the bacterial community gave risetoonly a few bands with low intensity in all treatments (Figure 1). Bacterial profile analysis of the highest remediation potential community (C) showed consistently greater band intensities (Figure 1). Bacterial communities in three different treatment samples represented three distinct clusters (Figure 1); cluster 1 was the smallest representing Week 0 samples of S (seawater + BH medium) and O (seawater + BH medium + weathered crude oil) and showed a similarity value of 67%. Cluster 2 represented treatment C (seawater + BH medium + weathered crude oil + consortia) for all samples from Week 0 to Week 4; Week 0 were different from Week 1; Week 2, Week 3 and Week 4 showed 66% similarity while Week 3 and Week 4 profile were84% similar (Figure 1). Cluster 3 represented Week 1 to Week 4 samples of O (seawater + BH medium + weathered crude oil) and Week 1 to Week 4 of S (seawater + BH medium) (Figure 1).

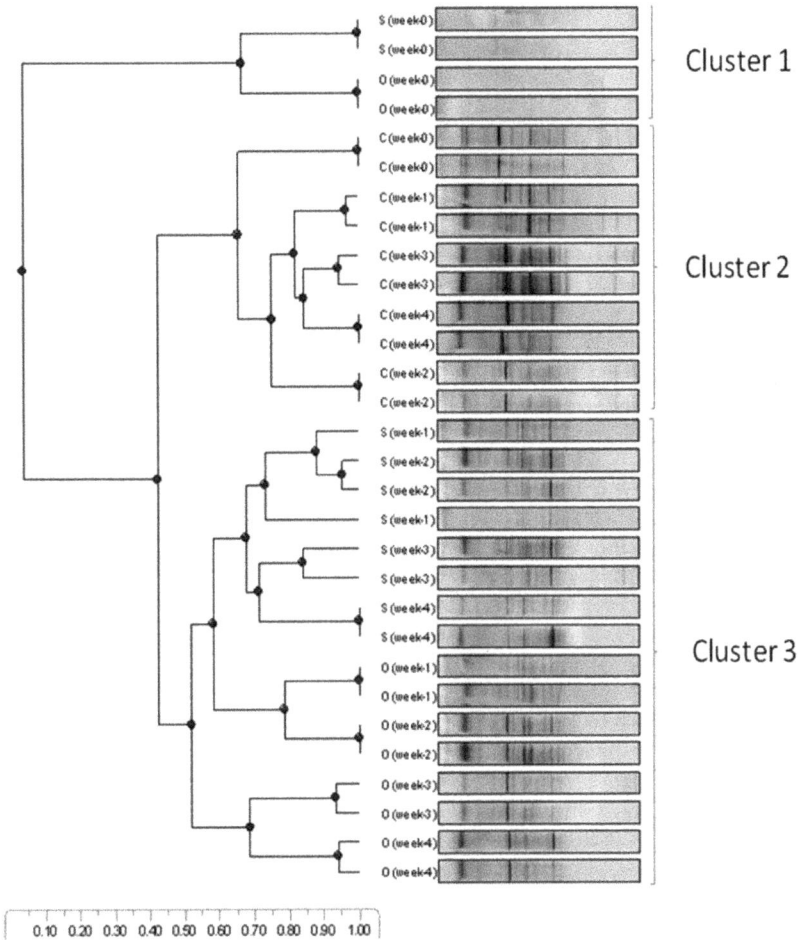

Figure 1. UPGMA dendrogram constructed from similarity matching data produced from the TGGE profiles of cDNA amplified from Week 0 to Week 4 of samples S (seawater + BH medium), O (seawater + BH medium + weathered crude oil), and C (seawater + BH medium + weathered crude oil + consortia). The scale bar represents percent similarity. Duplicate samples were analysed.

During the initial week (Week 0) the natural community in sea water should be similar in all treatments, so treatments S (seawater + BH medium), O (seawater + BH medium + weathered crude oil) from Week 0 represented the same cluster (Figure 1). However treatment C (seawater + BH medium + weathered crude oil + consortia) from Week 0 did not fall into same cluster (Cluster 1) presumably due to fact that the community in sea water was already changed by adding the consortia. Further weeks (Week 1 to Week 4) of treatment S (seawater + BH medium) formed the same cluster which included treatment O (sea water + BH medium + weathered crude oil) from Week 1 to Week 4, showing that the community in natural sea water changed from the initial week. Throughout the experiment, for Week 0 to Week 4 for treatment C (seawater + BH medium + weathered crude oil + consortia) all samples clustered together (Cluster 2) (Figure 1). However within the cluster, changes were observed, presumably due to changes in the dominance of the bacterial species. This was observed clearly in Week 3 (Figure 1).

It can be concluded that TGGE was found to be an excellent tool for analysing communities and separating those exhibiting a greater ability to carry out bioremediation from other, less active communities (Figure 1).

RNA-TGGE data was not only used to cluster the communities or differentiate the communities having high potential in bioremediation but can also be used for a range of analyses (Shannon Weaver diversity, Pareto-Lorenz curve) as well as for sequencing through excision of the TGGE bands which provides further information. All this data represents important management and reporting information relating to successful bioremediation.

3.2. Shannon Weaver Diversity

Shannon Weaver diversity represents an estimation of species richness [1,29]. Shannon Weaver diversity was calculated for each of the differently treated samples (C, O and S) (Figure 2). Greatest diversity was observed in sample C (seawater + BH medium+ weathered crude oil + consortia), with greatest diversity being observed in the Week 3 sample. Increased diversity is generally an indication of good resilience of a community and generally associated with relatively higher levels of bioremediation [30].

3.3. Pareto-Lorenz Curve

Pareto Lorenz curves represent the functional organisation of community. This organisation is the result of the action of microorganisms that are most fitting to the ongoing environmental-microbiological interactions [28]. Pareto-Lorenz curves were constructed using the band intensities of the three treatments S (seawater + BH medium), O (seawater + BH medium + weathered crude oil), and C (seawater + BH medium + weathered crude oil + consortia) representing Week 0 to Week 4. As a general rule, the more the PL curve deviates from the 45° diagonal (the theoretical perfect evenness line), the less evenness can be observed in the structure of the studied community. The latter means that a smaller fraction of different species is present in dominant numbers. The 25%, 45% and the 80% curves based on the Y-axis projection of their respective intercepts with the 20% X-axis represent low, medium and high functional organisation respectively [26,28].

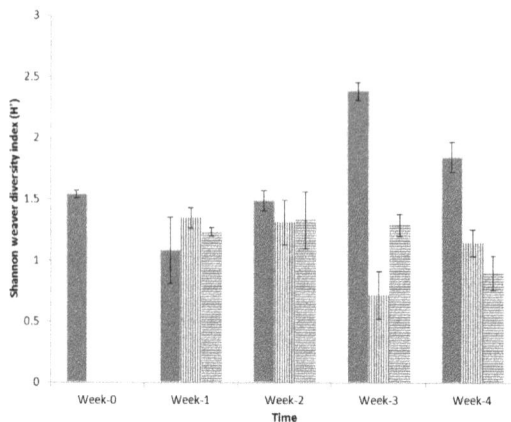

Figure 2. Diversity of three treatment samples "▬" C (seawater + BH medium + weathered crude oil + consortia); "|||" O (seawater + BH medium + weathered crude oil); and "☰" S (seawater + BH medium) from Week 0 to Week 4. Duplicate data (*n* = 2) were analysed.

Communities of treatment C (seawater + BH medium + weathered crude oil + consortia) in Week 2 and Week 3 suggested that the most active species (playing a major role in degradation) were dominant from the PL curve at 50% and 40% (Figure 3) [28]. The remaining Weeks (Week 0, Week 1 and Week 4) of treatment C (seawater + BH medium + weathered crude oil + consortia) represent a specialised community (PL curve around 60%) in which a small number of species were dominant and all others species were present in low numbers [31]. A similar specialised community was also observed in treatment O (Week 1 to Week 4), (seawater + BH medium + weathered crude oil) (Figure 4) and treatment S (seawater + BH medium) (Figure 5). This also explains why treatments O and S formed a cluster (Cluster 3) (Figure 1). There were no PL curves at Week 0 for both treatments (treatment O and treatment S) due to the low number of DNA bands detected (Cluster 1) (Figure 1).

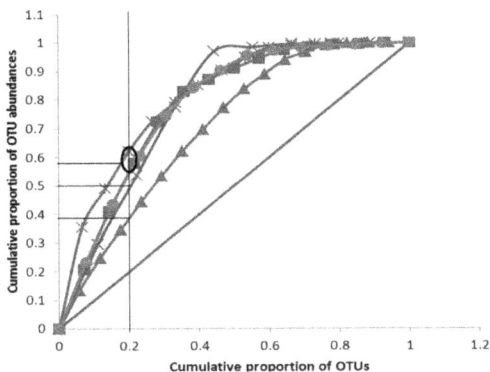

Figure 3. PL curves representing from Week 0 to Week 4 oftreatment C (seawater + BH medium + weathered crude oil + consortia). ("———" Perfect evenness line; "——■——" C (Week 4); "——▲——" C (Week 3); "——✸——" C (Week 2);"——✶——" C (Week 1);"——●——" C (Week 0)).Oval represents a grouping of curves (more than one) and horizontal line without oval represents asingle PL curve.

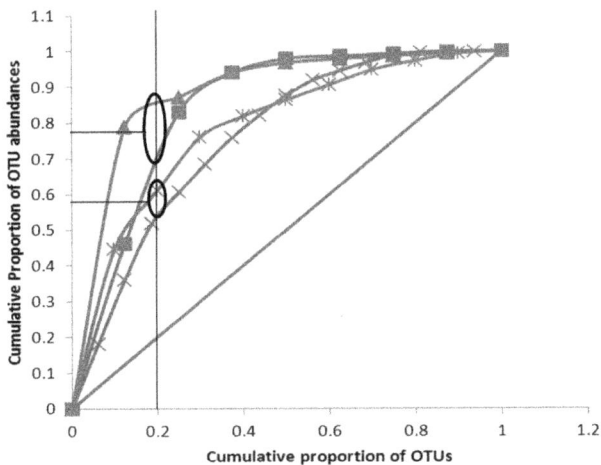

Figure 4. PL curves representing from Week 1 to Week 4 oftreatment O (seawater + BH medium + weathered crude oil). ("———" Perfect evenness line; "——■——" O (Week 4); "——▲——" O (Week 3); "——✸——" O (Week 2);"——✶——" O (Week 1)).

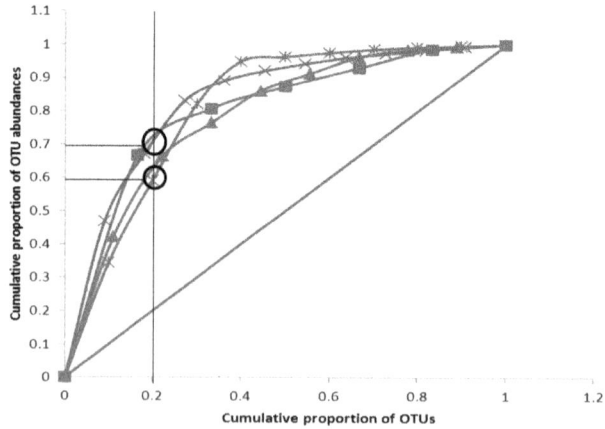

Figure 5. PL curves representing from Week 1 to Week 4 of treatment S (seawater + BH medium). ("━━━" Perfect evenness line; "━■━" S (Week 4); "━▲━" S (Week 3);"━✳━" S (Week 2); "━✳━" S (Week 1)).

3.4. RNA-TGGE Bands

The additional key feature of RNA-TGGE is that it allows to the excision and sequencing of bands from the gel, enabling the identification of members of the active community (Table 1). Sequenced data from TGGE excised bands showed the presence of two species in all the three treatments, namely a *Gammaproteobacteria* and an uncultured bacterium (Table 1). However, these species represented various identities (accession number) and similarities in each treatment confirming the fact that they are certainly unique in nature (Table 1).

Table 1. Different bacterial species identified from the TGGE gel bands excised in three different treatments (C, O and S), their identity and similarity.

Identified Species	Treatment	Accession No.	Similarity (%)
Pseudomonas sp. (*Gammaproteobacteria*)	C	JF778683.1	96
Uncultured marine bacterium	C	FM211075.1	93
Uncultured *Alcanivorax* sp. (*Gammaproteobacteria*)	C	JF979266.1	84
Uncultured bacterium clone	C	HQ827507.1	100
Uncultured organism clone	C	JN528201.1	88
Uncultured bacterium clone	C	HM580751.1	99
Uncultured bacterium clone	C	JN178389.1	100
Alcanivoraxborkuumensis (*Gammaproteobacteria*)	C	FJ218422.1	94
Alcanivorax sp. (*Gammaproteobacteria*)	C	AB681673.1	91
Alcanivoracaceae bacterium (*Gammaproteobacteria*)	O	HQ537302.1	97
Uncultured betaproteobacteruim	O	GQ274246.1	88
Alcanivorax sp. (*Gammaproteobacteria*)	O	HE586882.1	96
Uncultured bacterium	O	EU255846.1	81
Uncultured bacterium	S	DQ861039.1	91
Alcanivorax sp. (*Gammaproteobacteria*)	S	HM171217.1	95
Alcanivorax sp. (*Gammaproteobacteria*)	S	HM171217.1	98
Marinobactermobilis strain (*Gammaproteobacteria*)	S	NR_044456.1	95

In this present study RNA-TGGE was carried out using universal bacterial primers rather than specific primers, because the site to be remediated contains multiple types of compounds or contaminants [2,32] which is not ideal for the application of specific primers. In this study the contaminant was weathered crude oil, representing a complex mixture of tens of thousands of compounds [31,33–35]. However, RNA-TGGE can be readily applied to bioremediation where there are few contaminants through the use of specific gene primers.

Previous researchers have shown that TGGE is well suited for fingerprinting bacterial communities by separating PCR-amplified fragments [36] because it provides a crucial measurement for different habitants, thereby providing a comparative study [15]. In addition, this technique can be coupled with other techniques [13] so, this technique not only useful as a management tool but also can be useful technology in commercial bioremediation.

4. Conclusions

RNA-TGGE was found to be an excellent tool for analysing hydrocarbon-degrading bacterial communities in marine ecosystems representing a simple, reproducible, management tool for commercial bioremediation purposes.

Acknowledgments: This research was supported by the Australia India Strategic Research Fund (BFO20032) and also by the South Australian Premiers Science Research Fund.

Author Contributions: Conceived and designed the experiments: K.K.K., K.L.S. and A.S.B.; Performed the experiments: K.K.K., K.L.S. and P.J.S.; Analysed the data: K.K.K. and E.S.; Wrote the paper: K.K.K., E.S. and A.S.B.

Conflicts of Interest: The authors declare no conflict of interest.

References

1. Ogino, O.; Koshikawa, H.; Nakahara, T.; Uchiyama, H. Succession of microbial communities during a biostimulation process as evaluated by DGGE and clone library analysis. *J. Appl. Microbiol.* **2001**, *91*, 625–635. [CrossRef] [PubMed]
2. Kadali, K.K.; Simons, K.L.; Skuza, P.P.; Moore, R.B.; Ball, A.S. A complementary approach to identifying and assessing the remediation potential of hydrocarbonoclastic bacteria. *J. Microbiol. Methods* **2012**, *88*, 348–355. [CrossRef] [PubMed]
3. Roose-Amsaleg, C.L.; Garnier-Sillam, E.; Harry, M. Extraction and purification of microbial DNA from soil and sediment samples. *Appl. Soil Ecol.* **2001**, *18*, 47–60. [CrossRef]
4. Widada, J.; Nojiri, H.; Omori, T. Recent developments in molecular techniques for identification and monitoring of Xenobiotic-degrading bacteria and catabolic genes in bioremediation. *Appl. Microbiol. Biotechnol.* **2002**, *60*, 45–49. [PubMed]
5. Connon, S.A.; Giovannoni, S.J. High-throughput methods for culturing microorganisms in very-low-nutrient media yield diverse new marine isolates. *Appl. Environ. Microbiol.* **2002**, *68*, 3878–3885. [CrossRef] [PubMed]
6. Frias-Lopez, J.; Shi, Y.; Tyson, G.; Coleman, M.L.; Schuster, S.C.; Chisholm, S.W.; Delong, E.F. Microbial community gene expression in ocean surface waters. *Proc. Natl. Acad. Sci. USA* **2008**, *105*, 3805–3810. [CrossRef] [PubMed]
7. Muyzer, G.; de Waal, E.C.; Uitterlinden, A.G. Profiling of complex microbial populations by denaturing gradient gel electrophoresis analysis of polymerase chain reaction-amplified genes coding for 16S rRNA. *Appl. Environ. Microbiol.* **1993**, *59*, 695–700. [PubMed]
8. Nocker, A.; Burr, M.; Camper, A.K. Genotypic microbial community profiling: A critical technical review. *Microb. Ecol.* **2007**, *54*, 276–289. [CrossRef] [PubMed]
9. Muyzer, G. DGGE/TGGE a method for identifying genes from natural ecosystems. *Curr. Opin. Microbiol.* **1999**, *2*, 317–322. [CrossRef]
10. Muyzer, G.; Smalla, K. Application of denaturing gradient gel electrophoresis (DGGE) and temperature gradient gel electrophoresis (TGGE) in microbial ecology. *Antonie Van Leeuwenhoek* **1998**, *73*, 127–141. [CrossRef] [PubMed]
11. Alvarez, P.J.J.; Illman, W.A. *Bioremediation and Natural Attenuation*; John Wiley and Sons, Inc.: Hoboken, NJ, USA, 2006.
12. Felske, A.; Akkermans, A.D.L.; de vos, W.M. Quantification of 16S rRNAs in complex bacterial communities by multiple competitive reverse transcription-PCR in temperature gradient gel electrophoresis fingerprints. *Appl. Environ. Microbiol.* **1998**, *64*, 4581–4587. [PubMed]

13. Cheung, P.; Kinkle, B.K. *Mycobacterium* diversity and pyrine mineralization in petroleum-contaminated soils. *Appl. Environ. Microbiol.* **2001**, *67*, 2222–2229. [CrossRef] [PubMed]

14. Gomez-Villalba, B.; Calvo, C.; Vilchez, R.; Gonzalez-Lopez, J.; Rodelas, B. TGGE analysis of the diversity of ammonia-oxidizing and denitrifying bacteria in submerged filter biofilms for the treatment of urban wastewater. *Appl. Microbiol. Biotechnol.* **2006**, *72*, 393–400. [CrossRef] [PubMed]

15. Beier, S.; Witzel, K.; Marxsen, J. Bacterial community composition in central European running waters examined by temperature gradient gel electrophoresis and sequence analysis of 16S rRNA genes. *Appl. Environ. Microbiol.* **2008**, *74*, 188–199. [CrossRef] [PubMed]

16. Kaartokallio, H.; Tuomainen, J.; Kuosa, H.; Kuparinen, J.; Martikainen, P.J.; Servomaa, K. Succession of sea-ice bacterial communities in the Baltic sea fast ice. *Polar Biol.* **2008**, *31*, 783–793. [CrossRef]

17. Felske, A.; Wolterink, A.; Lis, R.V.; Akkermans, A.D.L. Phylogeny of the main bacterial 16S rRNA sequences in Drentse A grassland soils (The Netherlands). *Appl. Environ. Microbiol.* **1998**, *64*, 871–879. [PubMed]

18. Panas, P.; McMullan, G.; Dooley, J.S.G. RT-TGGE as a guide for the successful isolation of phosphonoacetate degrading bacteria. *J. Appl. Microbiol.* **2007**, *103*, 237–244. [CrossRef] [PubMed]

19. Nocker, A.; Camper, A.K. Novel approaches toward preferential detection of viable cells using nucleic acid amplification techniques. *FEMS Microbiol. Lett.* **2009**, *291*, 137–142. [CrossRef] [PubMed]

20. Wagner, R. The regulation of ribosomal RNA synthesis and bacterial cell growth. *Arch. Microbiol.* **1994**, *161*, 100–109. [CrossRef] [PubMed]

21. Boon, N.; Top, E.M.; Verstraete, W.; Siciliano, S.D. Bioaugmentation as a tool to protect the structure and function of an activated-sludge microbial community against a 3-chloroaniline shock load. *Appl. Environ. Microbiol.* **2003**, *69*, 1511–1520. [CrossRef] [PubMed]

22. Bustin, S.A.; Benes, V.; Nolan, T.; Pfaffl, M.W. Quantitative real-time RT-PCR—A perspective. *J. Mol. Endocrinol.* **2005**, *34*, 597–601. [CrossRef] [PubMed]

23. Smith, C.J.; Osborn, A.M. Advantages and limitations of quantitative PCR (Q-PCR)-based approaches in microbial ecology. *FEMS Microbiol. Ecol.* **2009**, *67*, 6–20. [CrossRef] [PubMed]

24. Zoetendal, E.G.; Akkermans, A.D.L.; de vos, W.M. Temperature gradient gel electrophoresis analysis of 16S rRNA from human fecal samples reveals stable and host-specific communities of active bacteria. *Appl. Environ. Microbiol.* **1998**, *64*, 3854–3859. [PubMed]

25. Cappello, S.; Denaro, R.; Genovese, M.; Giuliano, L.; Yakimov, M.M. Predominant growth of Alcanivorax during experiments on "oil spill bioremediation" in mesocosms. *Microbiol. Res.* **2007**, *162*, 185–190. [CrossRef] [PubMed]

26. Kadali, K.K.; Simons, K.L.; Sheppard, P.J.; Ball, A.S. Mineralisation of weathered crude oil by a hydrocarbonoclastic consortia in marine mesocosms. *Water Air Soil Pollut.* **2012**, *223*, 4283–4295. [CrossRef]

27. Girvan, M.S.; Bullimore, J.; Pretty, J.N.; Osborn, A.M.; Ball, A.S. Soil type is the primary determinant of the composition of the total and active bacterial communities in Arable soils. *Appl. Environ. Microbiol.* **2003**, *69*, 1800–1809. [CrossRef] [PubMed]

28. Marzorati, M.; Wittebolle, L.; Boon, N.; Daffonchio, D.; Verstraete, W. How to get more out of molecular fingerprints: Practical tools for microbial ecology. *Environ. Microbiol.* **2008**, *10*, 1571–1581. [CrossRef] [PubMed]

29. Ovreas, L. Population and community level approaches for analysing microbial diversity in natural environments. *Ecol. Lett.* **2000**, *3*, 236–251. [CrossRef]

30. Romantschuk, M.; Sarand, I.; Petanen, T.; Peltola, R.; Jonsson-Vihanne, M.; Koivula, T.; Yrjala, K.; Haahtela, K. Means to improve the effect of *in situ* bioremediation of contaminated soil: An overview of novel approaches. *Environ. Pollut.* **2000**, *107*, 179–185. [CrossRef]

31. Martinez-Gomez, C.; Vethaak, A.D.; Hylland, K.; Burgeot, T.; Kohler, A.; Lyons, B.P.; Thain, J.; Gubbins, M.J.; Davies, I.M. A guide to toxicity assessment and monitoring effects at lower levels of biological organization following marine oil spill in European waters. *ICES J. Mar. Sci.* **2010**, *67*, 1–14. [CrossRef]

32. Head, I.M.; Jones, D.M.; Roling, W.F.M. Marine microorganisms make a meal of oil. *Nat. Rev.* **2006**, *4*, 173–182. [CrossRef] [PubMed]

33. Onwurah, I.N.E.; Ogugua, V.N.; Onyike, N.B.; Ochonogor, A.E.; Otitoju, O.F. Crude oil spill in environment, effects and some innovative clean-up biotechnologies. *Int. J. Environ. Res.* **2007**, *1*, 307–320.

34. Rojo, F. Degradation of alkanes by bacteria. *Environ. Microbiol.* **2009**, *11*, 2477–2490. [CrossRef] [PubMed]

35. Yuste, L.; Corbella, M.E.; Turiegano, M.J.; Karlson, U.; Puyet, A.; Rojo, F. Characterization of bacterial strains able to degrade to grow on high molecular mass residues from crude oil processing. *FEMS Microbiol. Ecol.* **2000**, *32*, 69–75. [CrossRef] [PubMed]
36. Heuer, H.; Hartung, K.; Wieland, G.; Kramer, I.; Smalla, K. Polynucleotide probes that target a hypervariable region of 16S rRNA genes to identify bacterial isolates corresponding to bands of community fingerprints. *Appl. Environ. Microbiol.* **1999**, *65*, 1045–1049. [PubMed]

Journal of
Marine Science and Engineering

MDPI

Article

Oil Characterization and Distribution in Florida Estuary Sediments Following the *Deepwater Horizon* Spill

Mace G. Barron *, Jill Awkerman † and Sandy Raimondo †

U.S. Environmental Protection Agency, Gulf Ecology Division, 1 Sabine Island Drive,
Gulf Breeze, FL 32561, USA; Awkerman.jill@epa.gov (J.A.); Raimondo.sandy@epa.gov (S.R.)
* Author to whom correspondence should be addressed; barron.mace@epa.gov;
  Tel.: +1-850-934-9223; Fax: +1-850-934-2402.
† These authors contributed equally to this work.

Academic Editor: Merv Fingas
Received: 22 July 2015; Accepted: 17 September 2015; Published: 23 September 2015

Abstract: Barrier islands of Northwest Florida were heavily oiled during *the Deepwater Horizon* spill, but less is known about the impacts to the shorelines of the associated estuaries. Shoreline sediment oiling was investigated at 18 sites within the Pensacola Bay, Florida system prior to impact, during peak oiling, and post-wellhead capping. Only two locations closest to the Gulf of Mexico had elevated levels of total petroleum hydrocarbons (TPH) and total polycyclic aromatic hydrocarbons (PAHs). These samples showed a clear weathered crude oil signature, pattern of depletion of C9 to C19 alkanes and C0 to C4 naphthalenes, and geochemical biomarker ratios in concordance with weathered Macondo crude oil. All other locations and sample times showed only trace petroleum contamination. The results of this study are consistent with available satellite imagery and visual shoreline survey data showing heavy shoreline oiling limited to sandy beaches near the entrance to Pensacola Bay and shorelines of Santa Rosa Island.

Keywords: oil; geochemical biomarkers; *Deepwater Horizon*; Pensacola Bay; polycyclic aromatic hydrocarbons

---

1. Introduction

The *Deepwater Horizon* (DWH) oil rig exploded on 20 April 2010, initiating the discharge of 800 million liters of oil into the northern Gulf of Mexico over an approximately three month period [1–3]. The spill was the largest environmental disaster in United States history, and the largest accidental oil spill in human history [4]. Vast areas of the Gulf of Mexico were impacted by oil, including deep ocean, pelagic, and estuarine ecosystems. Over 20 million hectares of the Gulf of Mexico were closed to fishing and 1600 km of shoreline were visibly oiled [2,5]. Shoreline oiling was temporally and spatially heterogeneous, with the heaviest oiling occurring in coastal areas of eastern Louisiana, Mississippi, Alabama, and on the barrier islands of Northwest Florida [6]. Along the more heavily oiled sand beaches, some oil and sand mixed and accumulated in the nearshore subtidal zone resulting in formation of extensive submerged oil residue mats [7]. In Florida, Shoreline Cleanup and Assessment Technique (SCAT) surveys were focused on coastal areas, with only limited surveys performed within large estuaries, including Pensacola Bay [7].

Satellite imagery and nearshore trajectories showed oil in proximity to Pensacola Bay and potential impacts on Santa Rosa Island from 17 June to 3 July 2010 [5,8]. The heaviest oiling of Santa Rosa Island occurred on June 23, with all 60 km of the barrier island's southern shoreline impacted with visible free product and particulate oil. Near shore water and sediment samples from the area were reported to have elevated levels of TPH and PAHs [9]. Passive water sampling devices deployed by Allan *et al.* [10] at the entrance to Pensacola Bay showed elevated levels of bioavailable petrogenic PAHs in August and September 2010, but only background concentrations in May, June, and July 2010, and in spring 2011 follow up sampling. Anecdotal reports indicated that mousse, sheen, tar balls, and tar mats were present within the Pensacola Bay system for multiple weeks, with the first consistent reports beginning about 10 June 2010.

The objectives of the current study were to assess shoreline sediment oiling within the Pensacola Bay system during the DWH spill for comparison to coastal oiling observations. The Pensacola Bay system is a 370 km^2 low energy river-dominated estuarine system comprised of interconnected large bays, smaller tidal bayous, and Santa Rosa Sound located in Northwest Florida [11]. Sampling times occurred prior to visible shoreline oiling, during peak oiling, and following capping of the wellhead. Samples were analyzed for multiple petroleum related analytes, including polycyclic aromatic hydrocarbons (PAHs) and geochemical biomarkers. Samples were also analyzed for a range of other organic chemicals and metals to allow evaluation of the spatial heterogeneity of contamination relative to petrogenic chemicals.

2. Materials and Methods

2.1. Sample Collection

Eighteen sample locations were selected throughout the Pensacola Bay system on the basis of accessibility, with the objective of collecting at geospatially diverse areas that represented the major habitats of the open bay and bayou habitats (Table 1, Figure 1). Sample sites included the entrance to Pensacola Bay and locations ranging from approximately 43 km to the east and 36 km to the north, including three bayous on the western coast of the bay. Sites were selected to be accessible by automobile because many locations within the bay had limited boat access due to the presence of oil containment booms. Boom placements varied throughout the study period, but were consistent at the mouth of bayous and public beach areas near Pensacola Pass. Where booms were present, samples were taken on the outside of booms along the unprotected shoreline when possible (Table 1). Sites were sampled from mid-June through September during eight serial sampling events: 16 or 17 June, 24 or 25 June, 30 June or 1 July, 8 or 9 July, 22 or 23 July, 5 or 6 August, 18 or 19 August, and 29 or 30 September. Exceptions to the sampling regime include sites in East Bay and the Escambia River Delta, where sampling began on 25 June, and Naval Air Station where permission was obtained to begin sampling on 23 July 2010.

Samples were collected according to a quality assurance sampling plan. At each site, surficial sediments (2 to 5 cm) were collected using stainless steel spoons and placed in 1.8 L glass jars with Teflon lids and homogenized by mixing prior to storage. An additional 240 mL sample was similarly collected at each site to analyze for non-petroleum related contaminants. Samples were immediately placed in a cooler on ice and frozen when returned to the laboratory (-70 °C).

Table 1. A Shoreline sediment sample collection site locations and characteristics within Pensacola Bay, Florida, USA.

Site Name	Site [1] ID	Site [1] Number	Distance from Pass (km) [2]	Latitude	Longitude	Site Characteristics	Booms Present [3]
Fort Pickens	FP	1	1.1	30.3310	−87.2966	sandy beach	yes [4]
Santa Rosa Sound 1	SRS1	2	12.1	30.3273	−87.1823	sandy beach	No
Santa Rosa Sound 2	SRS2	3	17.7	30.3345	−87.1389	sandy beach	yes [5]
Santa Rosa Sound 3	SRS3	4	27.9	30.3531	−87.0414	sandy beach	yes [5]
Santa Rosa Sound 4	SRS4	5	43.2	30.3830	−86.8650	sandy beach	No
Santa Rosa Sound 5	SRS5	6	29.3	30.3852	−87.0135	sandy beach	No
Santa Rosa Sound 6	SRS6	7	22.1	30.3737	−87.0914	sandy beach	No
Naval Live Oaks S	NLOS	8	18.6	30.3641	−87.1276	sandy beach	No
Naval Live Oaks N	NLON	9	17.2	30.3696	−87.1426	sandy beach	No
East Bay	EB	10	24.4	30.3988	−87.0735	sandy beach	No
Escambia Riverdelta	ERD	11	36.2	30.5810	−87.1611	sand, organic mix	No
Scenic Bluffs	SB	12	21.2	30.4551	−87.1675	sand, organic mix	No
Bayou Texar south	BTS	13	15.4	30.4201	−87.1933	sand, silt, clay	yes [6]
Bayou Texar north	BTN	14	16.7	30.4315	−87.1902	sand, silt, clay	yes [6]
Bayou Chico east	BCE	15	11.9	30.4001	−87.2428	sand, silt, clay	No
Bayou Chico west	BCW	16	13.6	30.4037	−87.2604	sand, silt, clay	yes [6]
Bayou Grande	BG1	17	12.8	30.3762	−87.3031	sand, silt, clay	yes [6]
Naval Air Station	NAS	18	1.5	30.3441	−87.3072	sandy beach	yes [3]

[1] ID: Site identification site numbers shown on Figure 1; [2] Distance via waterways to Pensacola pass approximated as the most direct path in ArcGIS [12]; [3] Booms in place when sampling initiated through 5 August 2010; [4] Samples taken outside of boomed area; [5] Samples taken within boomed area; [6] Booms at mouth of bayou.

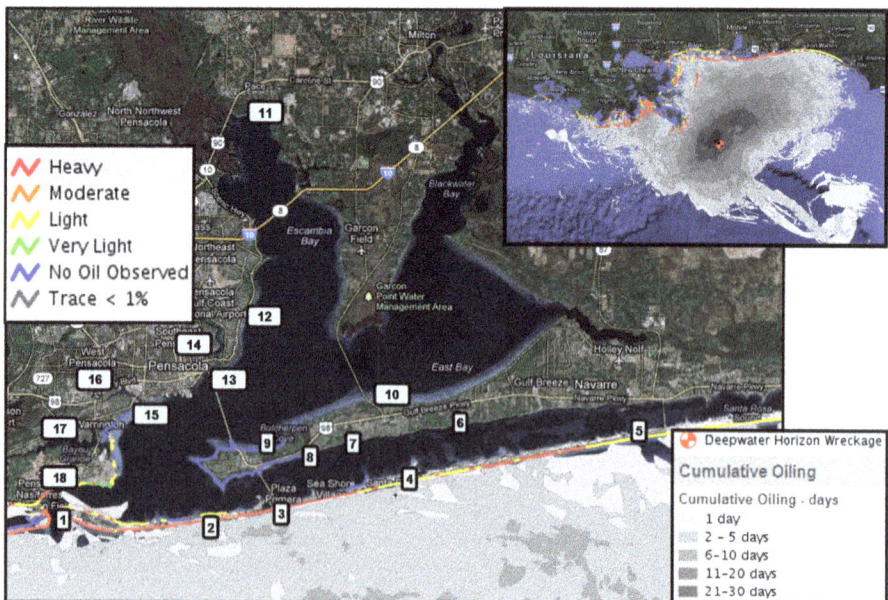

Figure 1. Composite graphic of Pensacola Bay, Florida derived from NOAA [6] data showing MODIS satellite imagery, maximum shoreline oiling (colored lines), and cumulative days of surface water oiling (grey surface shading). Sampling locations and identification numbers listed in Table 1. Inset: Gulf of Mexico, United States.

2.2. Analytical Chemistry

Sediment samples were extracted and analyzed for petroleum related analytes, metals, PCBs, and other organic contaminants (Tables S1–S4). Additionally, one sample of Macondo crude oil (MCO) collected by a remotely operated vehicle (ROV) directly at the wellhead was analyzed only for petrogenic chemicals. Sample holding, preservation, processing, and chemical analyses were performed following rigid chain of custody and quality assurance/quality control procedures according to USEPA methods and the Quality Assurance Project Plan of the contract laboratory. Sediment samples were mixed with sodium sulfate to remove moisture, than 20 g subsamples were prepared by automated Soxhlet extraction with dichloromethane followed by silica gel cleanup. Extracts analyzed for organochlorine pesticides had additional clean up by passing the extract through a Florisil column (elution with 10% acetone in hexane) and a solid phase carbon cartridge (elution with dichloromethane and hexane) to remove non-analyte interferences.

Petroleum-related analytes included total petroleum hydrocarbons (TPH), saturated hydrocarbons (SHC), petrogenic PAHs, and biomarkers. TPH representing the total aromatic and aliphatic hydrocarbon content of sample extracts were analyzed by gas chromatography/flame ionization detection (GC/FID) using a HP 5890 GC (Hewlett-Packard, Palo Alto, CA USA). Concentrations were determined from integration of the FID signal over the entire hydrocarbon range from n-C9 to n-C44 and were calibrated against an average alkane hydrocarbon response factor. Saturated hydrocarbons were analyzed by GC/FID based on EPA Method 8015 with the SHC fraction determined by integrating the resolved chromatographic peaks from the unresolved response. Individual alkanes including pristane, phytane, and C9 to C39 normal alkanes were quantified against a calibration curve made from C9 to C44 n-alkanes. Fifty-seven petroleum-related PAHs, including alkyl homolog groups, were analyzed by gas chromatography with mass spectrometry using selected ion monitoring (GC/MS-SIM) following the methods of Page *et al.* [13] and Wang and Stout [14]. The analytical procedure was based on EPA Method 8270D with the GC and MS operating conditions optimized for separation and sensitivity of the target analytes using an Agilent 5973 quadrupole GC/MS system (Agilent, Palo Alto, CA USA). Alkyl PAH homologs were quantified using a response factor assigned from the parent PAH compound. Fifty-six petroleum biomarkers were analyzed by GC/MS-SIM following the method of Wang *et al.* [15] using an Agilent 5973 quadrupole GC/MS system.

Non-petroleum analytes included heavy metals, PCBs, and pesticides. Seven metals were determined by inductively coupled plasma atomic emission spectrometry (ICP-AES) and inductively coupled plasma-mass spectrometry (ICP-MS) and total mercury by cold vapor atomic absorption using a CETAC M6200A mercury analyzer (CETAC Technologies, Omaha, NE USA). Total PCBs were analyzed by comparison to Aroclors following EPA method 8082. Twenty-six organochlorine pesticides were analyzed by GC/MS/MS with isotope dilution, including diphenyl, cyclodiene, and organophosphate insecticides following EPA method 1699 using a Waters Micromass Quattro Micro GC tandem MS (Waters, Milford, MA USA).

2.3. Data Analyses

Satellite imagery, cumulative surface water oiling, and shoreline oiling survey data for the Pensacola Bay area were downloaded from public domain databases [5] for comparison to analytical chemistry results. Weathering and diagnostic geochemical biomarker ratios were determined from the detectable concentrations of specific analytes following the equations in Table 2. Total PAH (tPAH) values were computed from the sum of detected analytes consisting of 57 parent PAHs and alkyl homolog groups. Depletion indices were computed from ratios of tPAH:hopane or the sum of C9 to C34 alkanes:hopane relative to MCO, and weathering ratios from C3 dibenzothiophenes: C3 chrysenes (Table 3).

Table 2. Total petroleum hydrocarbon (TPH) concentrations (mg/Kg) in shoreline sediment samples from Pensacola Bay, Florida during the 2010 *Deepwater Horizon* oil spill.

Sample Location [1]	Sediment TPH (mg/Kg) at Each 2010 Sample Date							
	16–17 Jun	24–25 Jun	30 Jun–1 Jul	8–9 Jul	22–23 Jul	5–6 Aug	18–19 Aug	29–30 Sept
1	6.65	197 *	366 *	4580 *	14.3	2.44	7.18	7.91
2	9.59	7.99	39.8 *	2.51	8.34	4.02	6.32	6.94
3	16.1	9.58	7.10	4.04	9.81	3.10	7.86	7.30
4	7.71	8.73	6.96	2.38	9.55	3.93	7.60	8.53
5	8.73	7.64	9.49	2.45	7.66	3.85	6.69	7.80
6	8.39	8.85	7.80	3.08	9.16	4.39	7.52	6.85
7	10.5	11.1	8.18	8.97	9.12	6.19	6.85	10.2
8	7.17	9.54	3.69	1.82	8.13	3.22	6.56	8.01
9	8.31	8.19	9.57	2.02	8.74	2.59	6.98	6.43
10	7.71	10.5	7.14	2.33	9.21	5.97	6.51	8.98
11	X [2]	9.56	13.6	2.83	10.70	4.70	8.60	13.6
12	X	8.33	9.39	2.89	9.80	4.15	9.56	10.6
13	35.7 *	27.2	26.6	21.0	35.4 *	23.7	21.0	45.7 *
14	13.9	11	20.1	18.5	16.4	12.2	14.9	22.8
15	23.2	19.3	12.8	9.54	13.8	8.82	12.1	36.4 *
16	12.2	24.3	18.5	52.2 *	11.5	8.47	10.9	18.7
17	11.5	9.94	8.43	2.62	8.29	4.94	7.54	11.4
18	X	X	X	X	18.3	5.62	105	13.7

[1] Locations are shown in Figure 1 and described in Table 1; [2] X: No sample collected because of restricted access; *Asterisk indicates values exceeding 30 ppm TPH

Table 3. Diagnostic geochemical biomarker ratios for Macondo source oil (MCO) and weathering in Pensacola Bay, Florida sediment samples (FP–July, NAS–August). [1]

Biomarker	Components	MCO	FP [1]	NAS [1]
Ts/Tm [2]	18α-22,29,30-trisnorneohopane/17α-22,29,30-trisnorhopane	1.28	1.20	0.911
Ts/(Ts+Tm) [3]	18α-22,29,30-trisnorneohopane/(18α-22,29,30-trisnorneohopane + 17α-22,29,30-trisnorhopane)	0.561	0.545	0.477
Triplet terp [2]	C26 tricyclic terpane 22S + 22R/C24 tetracyclic terpane	2.66	2.52	2.66
24Tri/23Tri [2]	C24 tricyclic terpane/C23 tricyclic terpane	0.508	0.798	0.812
26Tri/25Tri [2]	C26 tricylclic terpane 22S + 22R/C25 tricyclic terpane	1.03	1.06	NC [3]
28Tri/29Tri [2]	C28 tricylclic terpane 22S + 22R/C29 tricyclic terpane 22S + 22R	1.03	1.17	NC [3]
29D/29H [2]	18α(H)-30-norneohopane/17α(H),21β(H)-30-norhopane	0.401	0.398	0.349
C28R/C29R [2]	14α,17α-methylcholestane 20R/14α,17α-ethylcholestane 20R	1.00	0.987	1.139
C31S/(S+R) [4,5]	17α,21β-homohopane 22S/17α,21β-homohopane 22S + 22R	0.371	0.407	0.472
H29/H30 [2,6]	17α,21β-30-norhopane/17α,21β-hopane	0.493	0.427	0.466
Pri/Phy [6]	pristane/phytane	1.652	0.611	0.719
C31S/H30 [2]	17α,21β-homohopane 22S/17α,21β-hopane	0.227	0.266	0.325
C29S/(S+R) [6]	14α,17α-ethylcholestane 20S/14α,17α-ethylcholestane 20S + R	0.547	0.497	0.517
D2/P2 [2,7]	C2 dibenzothiophenes/C2 phenanthrenes	0.340	0.270	0.272
D3/P3 [2,7]	C3 dibenzothiophenes/C3 phenanthrenes	0.371	0.405	0.465
Pri/C17 [6]	pristane/n-C17	0.629	0.595	0.765
Phy/C18 [6]	phytane/n-C18	0.489	0.397	0.528
WR [8]	$\sum (nC23$ to $nC34)/\sum (nC11$ to $nC22)$	0.238	1.48 [3]	1.08 [3]
RPDI [9]	$[1 - (tPAH/hopane_{sample})/(tPAH/hopane_{oil})] \times 100$	0	99.5	98.9
RADI [9]	$[1 - (\sum alkanes/Hopane_{sample})/(\sum alkanes/Hopane_{source\ oil})] \times 100$	0	100	100
D3/C3 [7]	C3 dibenzothiophenes/C3 chrysenes	0.969	1.73	1.88

[1] Table 1 for site identifications and Figure 1 for sample location. Sediment samples collected July (FP) and August (NAS) 2010; [2] Rosenbauer *et al.* [3,16]; [3] One or more analytes below detection limits; [4] Mulabagal *et al.* [17]; [5] Hostettler *et al.* [18]; [6] Alkane to isoprenoid ratio [14]; [7] Douglas *et al.* [19]; [8] WR: Alkane weathering ratio; excludes pristane and phytane; [9] RPDI: Relative tPAH depletion index; RADI: Relative alkane depletion index. Modified from Atlas and Bragg [20].

3. Results

3.1. Oil Distribution

Satellite imagery and available cumulative surface water oiling data showed only trace oiling within Pensacola Bay (Figure 1) [6]. The limited SCAT survey (USGS 2011; NOAA 2013) data for the Pensacola Bay system during the DWH spill also indicated that oiling was restricted to areas of the outer bay (Figure 1) [6]. Although consistent repeated SCAT observations for these areas were not available, the reported maximum oiling levels indicated that heavy shoreline oiling was limited to sandy beach areas near the entrance to Pensacola Bay and the south shoreline of Santa Rosa Island. The 18 sample locations in Pensacola Bay had trace levels of TPH prior to observable shoreline oiling during the DWH spill (Table 2). Only the July 8 sample at Site 1 (FP) near the entrance to Pensacola Bay had very high elevations in both TPH (4580 mg/kg) and tPAH (13.2 mg/kg) (Table 2). Minor elevations in both TPH (52.2 mg/kg) and tPAH (0.3 mg/kg) were present at Site 16 (BCW) on July 9, and Site 18 (NAS) showed relatively high levels on Aug 19 (101 mg/kg TPH and 0.4 mg/kg tPAH). Site 13 (BTS; June 17) showed a minor elevation in pre-impact TPH (35.7 mg/kg) and tPAH (0.01 mg/kg); inspection of the specific analytes in that sample showed PAHs were elevated in pyrogenic, rather than petrogenic PAHs.

3.2. Weathering and Fingerprinting to Source Oil

Assessment of TPH levels, PAH and alkane distributions, and geochemical biomarker ratios in MCO and shoreline sediment samples collected prior to oil impact, during active oiling, and post well capping showed that only two sites had evidence of oil derived from the DWH spill. Samples collected during the period of active oiling at Fort Pickens (FP, Site 1) at the entrance to Pensacola Bay and Naval Air Station (NAS, Site 18) within 2 km of the entrance had elevated TPH and PAH concentrations, and norhopane, triterpane, and other biomarker ratios generally consistent with weathered MCO (Table 3). These samples also showed a characteristic pattern of depletion of C9 to C19 alkanes and C0 to C4 naphthalenes indicative of weathered crude oil, and higher concentrations of C2 and C3 PAH homologs (Figures 2 and 3). Relative tPAH and alkane depletion ratios indicated high weathering in both FP and NAS samples, whereas alkane to isoprenoid ratios were equivocal (Table 3). Pre-oiling and post well capping samples at FP and NAS, and all other sample locations and times had low concentrations of petroleum (Table 2, Figures 2 and 3).

3.3. Other Contaminants

Of the 72 non-petroleum analytes, 55 were not detected at any site (Table S2). None of the 26 organophosphates or seven PCB aroclors were detected. Of the 31 pesticides, 25 were not found at any of the sites; however, DDT isomers, dieldrin, and heptachlor epoxide were detected in Santa Rosa Sound (Sites 5, 6, and 7) as well as Bayou Texar (Sites 13 and 14), and Bayou Chico (Site 15; Table S3). Trace amounts of metals were found at most sites (Table S4). Bayous Chico and Texar had higher levels of metals than other sites. Lead levels in both bayous (Sites 14 and 15) and copper levels in north Santa Rosa Sound (site 7) were as much as 200-fold higher than levels detected at other sites (Table S4).

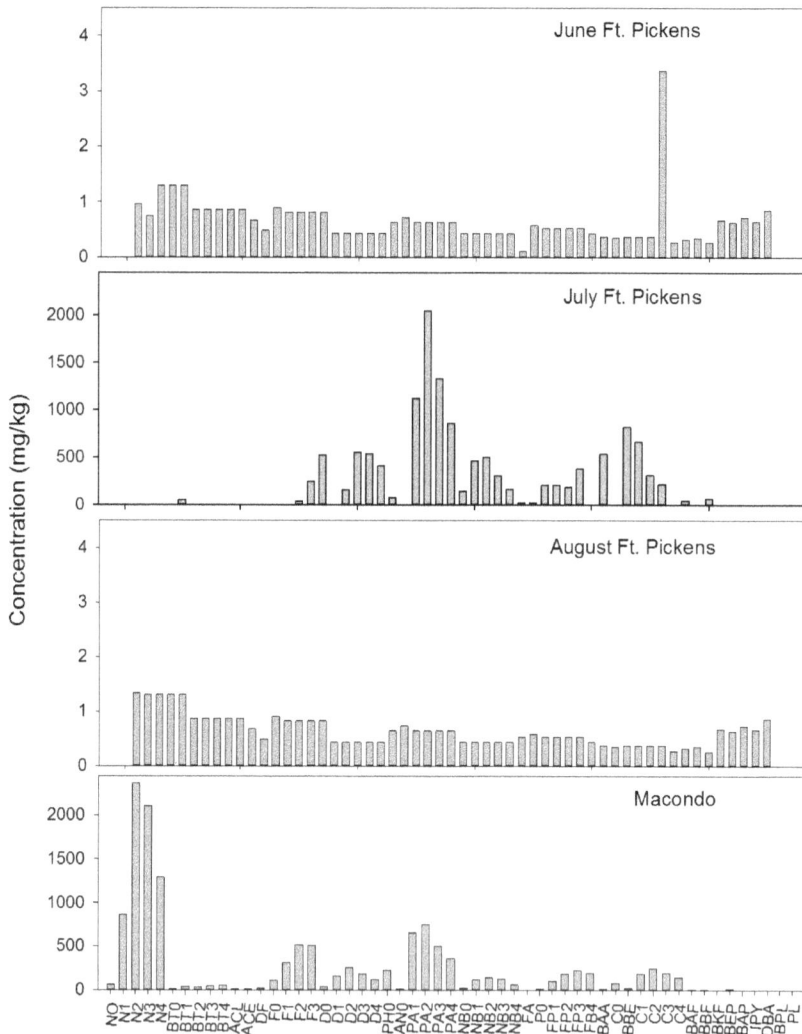

Figure 2. Composition of unsubstituted (0) and alkyl homologs (1 to 4 carbons) of PAHs in Macondo crude oil and shoreline sediment samples from Fort Pickens, Santa Rosa Island, Florida (Site 1; Figure 1, Table 1). Samples collected in June (Pensacola Beach pre-impact), July (following visible oiling), and August (post oiling). Note scale differences. N: napthalenes, BT: Benzothiophenes, ACL: Acenaphthylene; ACE: Acenaphthene; F: fluorenes; D: Dibenzothiophenes, PH: Phenanthrene; AN: Anthracene; PA: Phenanthrenes/anthracenes; NB: Napthobenzothiophenes; FA: Fluoranthene; PO: Pyrene; FP: Fluoranthenes/pyrenes; BAA: Benzo(a)anthracene; C: Chrysenes; BAF: Benzo(a)fluorene; BBF: Benzo(b)fluorene; BKF: Benzo(k)fluorene; BEP: Benzo(e)pyrene; BAP: Benzo(a)pyrene; IPY: Ideno(1,3,3-CD)perylene; DBA: Dibenzo(a,h)anthracene; BPL: Benzo(g,h,i)perylene; PL: Perylene.

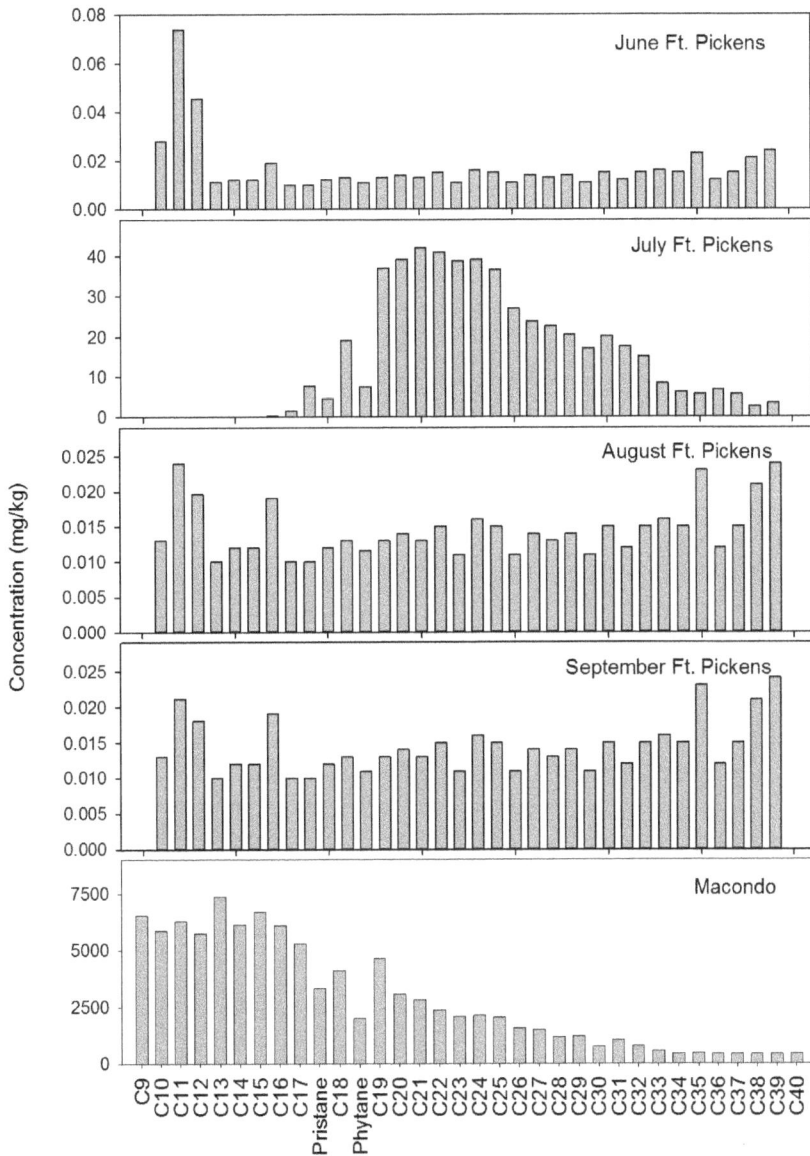

Figure 3. Alkane distribution (nonane, C9 to tetracontane, C40) in Macondo crude oil and shoreline sediment samples from Fort Pickens, Santa Rosa Island, Florida (Site 1; Figure 1, Table 1). Samples collected in June (Pensacola Beach pre-impact), July (during peak oiling), and August (post oiling). Note scale differences of samples.

4. Discussion

Over 1600 km of northern Gulf of Mexico shorelines were impacted from the DWH spill, with approximately 360 km heavily oiled [1,2,7]. SCAT survey results, satellite imagery, and cumulative oil determinations indicated that shoreline oiling was temporally and spatially heterogeneous [6].

In Florida, SCAT surveys were focused on the western barrier islands, with only a few surveys performed within the large estuary systems [6,7]. The analysis of petroleum analytes in the current study were consistent with available SCAT results and satellite imagery that heavy shoreline oiling within the Pensacola Bay system was limited to areas in proximity to the Gulf of Mexico. Other locations and sample times showed only trace petroleum contamination, and limited evidence of MCO. The June sampling dates preceded oil impacts on Santa Rosa Island and the Pensacola Bay system, consistent with sampling results in the current study. The single elevation in C2-chrysenes in the June Fort Pickens sample (Figure 2) appeared to be a minor anomaly compared to the 200 ppb of this specific PAH in the impacted July sample. Although visible oil occurred in proximity to Pensacola Bay for approximately two weeks, the hydrodynamics of the system may have limited more extensive shoreline oiling. Surface flows tend to be seaward, and based on average river flow and tidal range, the Pensacola Bay system should flush approximately every 34 days [11]. Additionally, the minimal tidal range of 0.5 m and primarily sandy shoreline sediments would tend to limit oil stranding and reduce residence time within the bay system.

Of the 138 sediment samples collected within the Pensacola Bay system in the current study and analyzed for TPH, only the July and August 2010 samples closest to the Gulf of Mexico had elevated levels of petroleum hydrocarbons and showed a clear signature of weathered MCO. Moderate weathering was indicated by the depletion of alkyl napthalenes and lower molecular weight alkanes, and similar concentrations of C2 and C3 PAH homologs [14]. Relative tPAH and alkane depletion, weathering ratios and quantitative biomarker ratios also were indicative of weathered MCO and consistent with other reported values [3,16,17,21]. For example, norhopane ratios (Ts/Tm; Ts/(Ts + Tm); 29D/29H) showed declines with distance from the Gulf of Mexico, whereas cyclic terpane (24Tri/23Tri) and hopane (C31S/(S + R); C31S/H30) ratios were elevated relative to MCO (Table 3). Triplet terp and cholestane (C28R/C29R) ratios were similar between MCO and the two impacted sites. Geochemical biomarkers have been used routinely in oil spill forensics since the *Exxon Valdez* incident because they are relatively resistant to degradation and oil formed under different geological conditions can have unique biomarker fingerprints [15,18]. Rosenbauer *et al.* [3,16] used a suite of diagnostic biomarkers to determine the presence of MCO oil in pre- and post- impact sediment and tar bar samples from Texas to Florida. The one Santa Rosa Island sample site (east of Navarre Beach, Florida) of Rosenbauer *et al.* [3,16] had no identifiable MCO in May 2010, whereas in October 2010 the sample results were indicative of a mixture of MCO and other oil sources. Mousse collected approximately 50 km west of the Santa Rosa Island Florida site in Alabama by Muglabagal *et al.* [17] during the June peak oiling period also showed a strong MCO signature.

Only the July Fort Pickens sample in the current study had high levels of TPH and petrogenic PAHs, and was the only sample to exceed screening level aquatic toxicity benchmarks for PAHs in sediment [22]. Maximum concentrations of 4600 mg TPH/kg were similar to levels reported by Kostka *et al.* [23] for the heavily exposed Gulf of Mexico side of Santa Rosa Island. These observations were consistent with the OSAT [24] report of a generally low incidence of coastal sediment samples exceeding aquatic toxicity benchmarks during the DWH spill. Sampling at Fort Pickens in August and September showed that petroleum contamination at this location had returned to pre-impact levels. OSAT [24] concluded that oil was weathering with variable degradation rates after the DWH spill, and bacterial gene sequencing revealed the presence of both alkane and PAH degraders in Santa Rosa island beach sand [23]. However, oil loss from the sandy shoreline sediments of the Pensacola Bay system may have been more dependent on tidal scouring and water washing than biodegradation. Remedial actions including beach cleaning and physical oil removal may also have contributed to oil declines [7]. Analysis of a diversity of potential other contaminants showed only minimal non-petroleum contamination of shoreline sediments within the Pensacola Bay system.

In contrast to petroleum contamination, metal and organic contaminants were largely restricted to anthropogenic source areas in proximity to boat docks and bayous, consistent with EPA [11] results for deeper surficial sediments. Alkyl homologs were absent in PAH profiles in shoreline sediment samples collected in areas of Pensacola Bay distant from the Gulf of Mexico and were indicative of combustion sources rather than oiling. The conclusion that *Deepwater Horizon* oiling of shoreline sediments was limited within the Pensacola Bay must be considered in the context of the sampling design. Sampling focused on surficial sediments of the estuary system and did not target deeper sediments, submerged tar mats, or the heavily oiled southern shoreline sediments of Santa Rosa Island within the Gulf of Mexico, areas which are known to be impacted by the spill [25]. Additional research and analysis of historical samples would be needed to address areas not targeted in this study.

Acknowledgments: Thanks to Greg Salata and staff at ALS for analytical chemistry, to Robyn Conmy for review of a draft of the manuscript, and to Becky Hemmer, Crystal Jackson, Hannah Rutter, Michael Norberg, and Alex Almario for field team assistance. The opinions expressed in this work are those of the authors and do not represent the policies or opinions of the U.S. EPA.

Author Contributions: M.G.B., J.A. and S.R. conceived of the study and designed the sampling and analysis plan. J.A. and S.R. performed sediment collections. J.A. and M.G.B. compiled, summarized, and interpreted the results. M.G.B. drafted the manuscript. All authors read and approved the manuscript.

Conflicts of Interest: The authors declare no conflict of interest.

References

1. USCG. On Scene *Coordinator Report Deepwater Horizon Oil Spill*. Submitted to National Response Team September 2011. United States Coast Guard: Cincinnati, OH, USA, 2010. Available online: http://www.uscg.mil/foia/docs/dwh/fosc_dwh_report.pdf (accessed on 15 January 2015).
2. Barron, M.G. Ecological impacts of the Deepwater Horizon oil spill: Implications for immuntoxicity. *Toxicol. Pathol.* **2012**, *40*, 315–320. [CrossRef] [PubMed]
3. Rosenbauer, R.J.; Campbell, P.L.; Lam, A.; Lorenson, T.D.; Hostettler, F.D.; Thomas, B.; Wong, F.L. *Reconnaissance of Macondo-1 Well Oil in Sediment and Tarballs from the Northern Gulf of Mexico Shoreline, Texas to Florida*; U.S. Geological Survey Open-File Report 2010–1290; United States Coast Guard: Cincinnati, OH, USA, 2010. Available online: http://pubs.usgs.gov/of/2010/1290/of2010-1290.pdf (accessed on 15 January 2015).
4. Eckle, P.; Burgherr, P.; Michaux, E. Risk of large oil spills: A statistical analysis in the aftermath of Deepwater Horizon. *Environ. Sci. Technol.* **2012**, *46*, 13002–13008. [CrossRef] [PubMed]
5. Carriger, J.; Barron, M.G. Minimizing risks from spilled oil to ecosystem services using influence diagrams: The Deepwater Horizon spill response. *Environ. Sci. Technol.* **2011**, *45*, 7631–7639. [CrossRef] [PubMed]
6. NOAA. *Environmental Response Management Application (ERMA) Deepwater Gulf Response. National Oceanographic and Atmospheric Administration*; NOAA: Washington, DC, USA, 2013. Available online: http://resources.geoplatform.gov/news/mapping-response-bp-oil-spill-gulf-mexico (accessed on 25 September 2013).
7. Michel, J.; Owens, E.H.; Zengel, S.; Graham, A.; Nixon, Z.; Allard, T.; Holton, W.; Reimer, P.D.; Lamarche, A.; White, M.; *et al.* Extent and degree of shoreline oiling: Deepwater Horizon oil spill, Gulf of Mexico, USA. *PLoS ONE* **2013**, *8*, e65087. [CrossRef] [PubMed]
8. Dietrich, J.C.; Trahan, C.J.; Howard, M.T.; Fleming, J.G.; Weaver, R.J.; Tanaka, S.; Yu, L.; Luettich, R.A.; Dawson, C.N.; Westerink, J.J.; *et al.* Surface trajectories of oil transport along the northern coastline of the Gulf of Mexico. *Cont. Shelf Res.* **2012**, *41*, 17–48. [CrossRef]
9. Sammarco, P.W.; Kolian, S.R.; Warby, R.A.F.; Bouldin, J.L.; Subra, W.A.; Porter, S.A. Distribution and concentrations of petroleum hydrocarbons associated with the BP/Deepwater Horizon oil spill, Gulf of Mexico. *Mar. Pollut. Bull.* **2013**, *73*, 129–143. [CrossRef] [PubMed]
10. Allan, S.E.; Smith, B.W.; Anderson, K.A. Impact of Deepwater Horizon oil spill on bioavailable polycyclic aromatic hydrocarbons in Gulf of Mexico coastal waters. *Environ. Sci. Technol.* **2012**, *46*, 2033–2039. [CrossRef] [PubMed]
11. USEPA. *The Ecological Condition of the Pensacola Bay System, Northwest Florida*; EPA/620/R-05/002; United States Environmental Protection Agency, Office of Research and Development: Washington, DC, USA, 2005.
12. ESRI. *ArcGIS Desktop: Release 10*; Environmental Systems Research Institute: Redlands, CA, USA, 2011.

13. Page, D.S.; Boehm, P.D.; Douglas, G.S.; Bence, A.E. Identification of hydrocarbon sources in the benthic sediments of Prince William Sound and the Gulf of Alaska following the Exxon Valdez oil spill. In *Exxon Valdez Oil Spill: Fate and Effects in Alaskan Waters*; ASTM STP 1219; Wells, P.G., Bulter, J.N., Hughes, J.S, Eds.; American Society for Testing and Materials: Philadelphia, PA, USA, 2008; pp. 44–83.

14. Wang, Z.; Stout, S.A. Chemical fingerprinting of spilled or discharged petroleum—Methods and factors affecting petroleum fingerprints in the environment. In *Oil Spill Environmental Forensics: Fingerprinting and Source Identification*; Wang, Z., Stout, S.A., Eds.; Elsevier Publishing Company: Boston, MA, USA, 2007; pp. 1–53.

15. Wang, Z.; Stout, S.A.; Fingas, M. Forensic fingerprinting of biomarkers for oil spill characterization and source identification. *Environ. Forensics* **2006**, *7*, 105–146. [CrossRef]

16. Rosenbauer, R.J.; Campbell, P.L.; Lam, A.; Lorenson, T.D.; Hostettler, F.D.; Thomas, B.; Wong, F.L. *Petroleum Hydrocarbons in Sediment from Northern Gulf of Mexico Shoreline, Texas to Florida*; U.S. Geological Survey Open-File Report 2011-1014; USGS: Reston, VA, USA, 2011. Available online: http://pubs.usgs.gov/of/2011/1014/of2011-1014.pdf (accessed on 15 January 2015).

17. Mulabagal, V.; Yin, F.; John, G.F.; Hayworth, J.S.; Clement, T.P. Chemical fingerprinting of petroleum biomarkers in Deepwater Horizon oil spill samples collected from Alabama shoreline. *Mar. Pollut. Bull.* **2013**, *70*, 147–154. [CrossRef] [PubMed]

18. Hostettler, F.D.; Lorenson, T.D.; Bekins, B.A. Petroleum fingerprinting with organic markers. *Environ. Forensics* **2013**, *14*, 262–277. [CrossRef]

19. Douglas, G.S.; Bence, A.E.; Prince, R.C.; McMillen, S.J.; Buttler, E.L. Environmental stability of selected petroleum hydrocarbon source and weathering ratios. *Environ. Sci. Technol.* **1996**, *30*, 2332–2339. [CrossRef]

20. Atlas, R.M.; Bragg, J.R. Bioremediation of marine oil spills: When and when not—The Exxon Valdez experience. *Microbiol. Biotech.* **2009**, *2*, 213–221.

21. Schantz, M.M.; Kucklick, J.R. *Interlaboratory Analytical Comparison Study to Support Deepwater Horizon Natural Resource Damage Assessment: Description and Results for Crude Oil QA10OIL01*; NISTIR 7793; National Institute of Standards and Technology: Gaithersburg, MD, USA, 2011.

22. Buchman, M.F. *NOAA Screening Quick Reference Tables*; NOAA OR&R Report 08-1; National Oceanic and Atmospheric Administration: Seattle, WA, USA, 2008; p. 34.

23. Kostka, J.E.; Prakash, O.; Overholt, W.A.; Green, S.J.; Freyer, G.; Canion, A.; Delgardio, J.; Norton, N.; Hazen, T.C.; Huettel, M. Hydrocarbon-degrading bacteria and the bacterial community response in Gulf of Mexico beach sands impacted by the Deepwater Horizon Spill. *Appl. Environ. Microbiol.* **2011**, *77*, 7962–7974. [CrossRef] [PubMed]

24. OSAT. *Summary Report for Fate and Effects of Remnant Oil Remaining in the Beach Environment*; Operational Science Advisory Team (OSAT-2); United States Coast Guard: Cincinnati, OH, USA, 2011. Available online: http://www.restorethegulf.gov/sites/default/files/u316/OSAT-2%20Report%20no%20ltr.pdf (accessed on 15 January 2015).

25. Hayworth, J.S.; Clement, T.P.; Valentine, J.F. Deepwater Horizon oil spill impacts on Alabama beaches. *Hydrol. Earth Syst. Sci.* **2011**, *15*, 3639–3649. [CrossRef]

Journal of
Marine Science and Engineering

MDPI

Article

Longer-Term Mental and Behavioral Health Effects of the Deepwater Horizon Gulf Oil Spill

Tonya Cross Hansel *, Howard J. Osofsky, Joy D. Osofsky and Anthony Speier

Department of Psychiatry, Louisiana State University Health Sciences Center, 1542 Tulane Avenue, New Orleans, LA 70433, USA; HOsofs@lsuhsc.edu (H.J.O.); JOsofs@lsuhsc.edu (J.D.O.); Aspei1@lsuhsc.edu (A.S.)

* Author to whom correspondence should be addressed; tcros1@lsuhsc.edu; Tel.: +1-504-568-6004.

Academic Editor: Merv Fingas
Received: 7 August 2015; Accepted: 12 October 2015; Published: 20 October 2015

Abstract: Mental health issues are a significant concern after technological disasters such as the 2010 Gulf Oil Spill; however, there is limited knowledge about the long-term effects of oil spills. The study was part of a larger research effort to improve understanding of the mental and behavioral health effects of the Deepwater Horizon Gulf Oil Spill. Data were collected immediately following the spill and the same individuals were resampled again after the second anniversary ($n = 314$). The results show that mental health symptoms of depression, serious mental illness and posttraumatic stress have not statistically decreased, and anxiety symptoms were statistically equivalent to immediate symptoms. Results also showed that the greatest effect on anxiety is related to the extent of disruption to participants' lives, work, family, and social engagement. This study supports lessons learned following the Exxon Valdez spill suggesting that mental health effects are long term and recovery is slow. Elevated symptoms indicate the continued need for mental health services, especially for individuals with high levels of disruption resulting in increased anxiety. Findings also suggest that the longer-term recovery trajectories following the Deepwater Horizon Gulf Oil Spill do not fall within traditional disaster recovery timelines.

Keywords: behavioral modifications; oil spill; anxiety

1. Introduction

Existing research suggests a number of negative mental health consequences for communities directly affected by oil spills [1]. In a community survey carried out in 1989, one year after the Exxon Valdez oil spill, Palinkas, Petterson, Russell and Downs [2] found a significant increase in rates of anxiety, posttraumatic stress disorder, and depression in residents with a high level of exposure to the spill and subsequent cleanup efforts. They also found a relationship between exposure to the oil spill and increased alcohol and substance use, domestic violence, chronic physical conditions, and a decline in social relationships. Those most vulnerable were groups with significant exposure and dependence on fishing and oil work for subsistence [3]. In an earlier study of the Sea Empress Oil Spill in Wales [4], the social and economic consequences following the spill resulted in increased concerns about health, finances, and perceived environmental risk; all of these factors resulted in increases in mental health symptoms [5]. Greater exposure resulting in increased behavioral health symptoms was also evident in the research done after the 2002 Prestige Oil Spill in Spain [6,7]. While several earlier studies of behavioral health following oil spills suggest an immediate negative impact, few studies explore the longer-term effects following oil spills.

1.1. Longer Term Effects of Oil Spills

Because most of the research concerning mental health effects following oil spills has been conducted within one year of the spill, there is a significant gap in the literature on how communities respond to the continued stress and changing environment following oil spills. Studies following the Exxon Valdez oil spill provide limited understanding of long-term mental and behavioral health effects indicating that the impact of oil spills persists for extended periods of time [8,9]. Eight years after the spill, Picou and Arata [10] found elevated levels of depression, intrusive stress, avoidance, and family conflict. Lessons learned from the Exxon Valdez spill show that individual and community effects lasted for decades, with at least part of the fishing industry unable to completely recover. In addition, destruction of the ecosystem occurs with oil spills that impacts on individuals and communities dependent on natural resources for their social and economic livelihood [11], thus, disrupting the usual networks of support that communities depend on to cope with adversities. With loss of jobs and livelihood, families may have few choices; they either have to move or live apart [3,9,12].

1.2. Longer-Term Disaster Recovery

There is limited research on longer-term mental health outcomes following oil spills, however findings increased mental health concerns for almost a decade following Exxon Valdez [10], suggest reevaluation of national disaster recovery timelines [13]. The Substance Abuse and Mental Health Services Administration's (SAMHSA) Disaster Kit suggests that the initial expected response and recovery trajectory focuses on the phases of heroism, honeymoon, and disillusionment, with reconstruction and the new beginning coinciding with the first year anniversary [14]. The surge in initial recovery efforts is often remarkable within the first year providing the boost needed for individuals, families, and communities to begin to move forward with the more prolonged recovery tasks. In most instances, the vast majority of those impacted have dealt with their recovery requirements within 12–18 months after the incident [15]. The 18-month timeline for disaster recovery is also evident in the Federal Emergency Management Agency (FEMA) recovery work timeline, where the final date for permanent work ends at 18 months, marking the end of recovery [16]. However, FEMA also states that time extensions may be granted for complicated disasters [16]. Depending on disaster experiences, personal history and recovery environment, behavioral health effects can linger far beyond the physical recovery and cleanup.

In addition, the Centers for Disease Control have noted that the phases and timelines of disaster recovery have been observed and developed based on natural disasters [15]. For natural disasters, studies attempting to understand the longer-term mental and behavioral health consequences are varied. In a recent review article, MacFarlane and Williams [15] noted anxiety disorder rates ranging from 2% to 29% in longitudinal studies. While many disaster studies report a natural remission [15], population studies have shown that diagnosis of PTSD can be chronic and take upwards of 72 months to remit [17]. Specific to the Gulf South, rates of probable PTSD remained elevated two years following Hurricane Katrina with over 40% endorsing symptoms [18]. The variance in rates and length is due to many factors including sampling, longevity of disaster (*i.e.*, whether it had a clear beginning or ending), magnitude of disaster, preparedness, co-morbidity, and subsequent traumas. Clearly there is a need for more research understanding the longer-term recovery trajectories following all disasters and specifically for technological disasters. Given the historical presence of disasters along the Gulf Coast some individuals may remain in the stage of disillusionment as recovery becomes increasingly more elusive. This outcome seems to have occurred following the Deepwater Horizon Incident otherwise known as the Gulf Oil Spill, where environmental, ecological, and economic effects of the oil spill are still largely unknown.

2. Deepwater Horizon Gulf Oil Spill

The Deepwater Horizon (DWH) Gulf Oil Spill, caused by an offshore oil platform explosion about 50 miles southeast of the Mississippi River delta, occurred on 20 April 2010. Deepwater Horizon spewed an estimated five million barrels of oil for three consecutive months, and is the largest marine oil spill in history [11]. Given the uniqueness of the spill, especially its size and occurrence less than five years following the worst natural disaster in United States' history, Hurricane Katrina, it is difficult to make assumptions about the impact on areas affected.

The Louisiana State University Health Sciences Center Department of Psychiatry conducted a study designed to assess the immediate mental health impact on residents in Southeastern Louisiana heavily impacted by the Gulf Oil Spill using telephone and face-to-face interviews. The results showed that the factors having the greatest effect on mental health were the extent of disruption on participants' lives, work, family, and social engagement resulting in increased symptoms of anxiety, depression, and posttraumatic stress. Given that the location of the oil spill affected individuals and communities with prior devastation from Hurricane Katrina, results also revealed that losses from Hurricane Katrina were highly associated with negative mental health outcomes, however the oil spill distress had unique variance in the analyses supporting that the DWH Gulf Oil Spill represents a complex recovery [19]. Additional studies conducted across the Gulf States have concurred with these findings and support the need for continued mental health treatment of children and adults, due to increased mental health concerns and symptoms [20–24]. In contrast, findings from a federal studies found a lack of increase in mental health symptoms following the oil spill; however, the authors note that a limitation with their study is that the broad population based surveillance methods may underestimate prevalence due to individuals directly affected living in smaller sub-communities [25].

The DWH Gulf Oil Spill studies demonstrate the immediate mental and behavioral health impact and subsequent needs following the disaster. Based on both clinical experience and supportive work done in communities along the Louisiana Gulf Coast, the current study hypothesized that negative mental health symptoms would remain elevated longer than the traditional one-year disaster recovery timeline. Consistent with disaster literature, it was hypothesized further that continued symptomatology would be associated with greater perceived disruption from the DWH Gulf Oil Spill. This study aims to explore recovery of a sample of Gulf Coast residents assessed in the first year following the spill and again just after the second Anniversary. The overall goal is to improve understanding of the longer-term impact of oil spills.

3. Experimental Design

This study was part of a larger research effort designed to improve understanding of the mental and behavioral health effects on individuals following the DWH Gulf Oil Spill. The first set of data was collected 1 year following the spill (Time 1) and the second set was gathered one year later after the second anniversary (Time 2). Time 1 began in August 2010 and with funding provided by the Louisiana Department of Social Services and ended in October 2011 with funding from the Louisiana State Department of Health and Human Services, Office of Behavioral Health. Coinciding with changes in funding and to increase the comprehensiveness of symptoms assessed, additional measures assessing depression and anxiety were added mid surveillance in Time 1. A total of 2093 participants were surveyed in Time 1 using both random telephone and purposive sampling. Participants from Time 1 were resampled following the second anniversary of the spill beginning in April 2012 and ending in August 2012. Interviews were conducted over the telephone using valid numbers provided in Time 1. Three attempts were made to contact each person by telephone and a total of 769 successful contacts were made. Of those contacted, a total sample of 314 agreed to participate, were matched based on last name and birthdate, and provided valid responses. The minimum time between surveys was 5 months and the maximum was 22 months ($M = 13.89$, $SD = 4.76$). The research protocol was approved by the Louisiana State University Health Sciences Center Institutional Review board.

4. Measures

The Deepwater Horizon Psychosocial Assessment was developed with consultation from stakeholders, local leaders, and state and national consultants. The assessment was comprised of the following sections measuring: socio-demographics, Hurricane Katrina losses, oil spill concerns and disruption, and mental health.

Hurricane Katrina experiences: Respondents were asked to endorse if they had experienced the following as a result of Hurricane Katrina in 2005: house destroyed, house damaged, injured, loss of business, loss of income, family members injured, family members killed, loss of personal property other than house, became seriously ill, victimized, friends/family members house destroyed/damaged, friends injured, and friends killed. A Hurricane Katrina experience index was created where 1 point was given for endorsement of each variable. The minimum score was 0 and the maximum was 11 ($M = 4.04$, $SD = 2.38$).

Oil spill concerns and disruption: Respondents were asked to endorse if they had concerns or problems with the following as a result of the DWH Gulf Oil Spill: damage to wildlife and environment; health and food concerns; loss of usual way of life; loss of job opportunities; loss of tourism; personal health effects; loss of personal or family business; and needing to relocate. An oil spill concern index was created where 1 point was given for endorsement of each variable. The minimum score was 0 and the maximum was 8 (Time 1, $M = 4.64$, $SD = 2.26$; Time 2, $M = 4.69$, $SD = 2.42$). A modified version of the Sheehan Disability Scale (SDS) was used to assess overall disruption of life from the oil spill [26]. Participants were asked to rate the extent to which the oil spill disrupted their work, school work, social life and leisure activities, and family life and home responsibilities on a five-point Likert scale ranging from 1 (*not at all*) to 5 (*extremely*). The minimum score was 3 and the maximum was 15 (Time 1, $M = 7.92$, $SD = 4.21$; Time 2, $M = 7.22$, $SD = 4.10$).

Mental health: Mental health was assessed using the K6 [27], Posttraumatic Symptom Checklist for Civilians (PCL-C) [28], Center for Epidemiologic Studies Depression Scale (CESD) [29], and General Anxiety Disorder (GAD-7) [30].

The K6 was used to assess overall well-being and, specifically, symptoms related to anxiety and depression. Respondents were asked to rate on a five-point Likert scale ranging from 0 (*none of the time*) to 4 (*all of the time*) if they felt: nervous, hopeless, restless or fidgety, so depressed that nothing could cheer them up, that everything was an effort, and if they felt worthless. Scores range from 0 to 24 and the minimum score for the current sample was 0 and the maximum was 24 (Time 1, $M = 6.80$, $SD = 6.61$, $\alpha = 0.94$; Time 2, $M = 6.19$, $SD = 6.56$, $\alpha = 0.92$). A cut-off score of 13+ was used to determine significant symptoms of serious mental illness; 63 (21%) met the cut off at Time 1 and 62 (20%) at Time 2.

Posttraumatic stress symptoms were assessed using the PCL-C. The 17 item scores range from 1 (*not at all*) to 5 (*extremely*) and total scores can range from 17 to 85. The minimum score for the current sample was 17 and the maximum was 85 (Time 1, $M = 34.93$, $SD = 16.80$, $\alpha = 0.97$; Time 2, $M = 34.45$, $SD = 18.42$, $\alpha = 0.96$). A cut-off score of 50 was used to determine significant symptoms of posttraumatic stress; 59 (20%) met the cut off at Time 1 and 66 (21%) at Time 2.

Depression was measured using the CESD. The 10 item scores are assigned values from 0 (*none of the time*) to 3 *all of the time* and total score ranges from 0 to 30. The minimum score for the current sample was 0 and the maximum was 30 (Time 1, $M = 9.60$, $SD = 9.32$, $\alpha = 0.94$; Time 2, $M = 9.06$, $SD = 9.13$, $\alpha = 0.95$). A cut-off score of 10+ was used. At Time 1 ($n = 172$), 73 (42%) met the cut off and at Time 2 ($n = 313$), 53 (17%) met the cut off for depression.

Anxiety was measured using the GAD-7. The 7 item scores are assigned values from 0 (*not at all*) to 3 (*nearly every day*); total score for the 7 items ranges from 0 to 21. Scores of 5, 10, and 15 are taken as the cut off points for mild, moderate, and severe anxiety, respectively. The minimum score for the current sample was 0 and the maximum was 21 (Time 1, $M = 7.97$, $SD = 7.21$, $\alpha = 0.96$; Time 2, $M = 7.40$, $SD = 7.06$, $\alpha = 0.94$). At Time 1 ($n = 172$), 44 (26%) met the cut off for mild anxiety, 21 (12%) for moderate anxiety and 37 (22%) for severe anxiety. At Time 2 ($n = 314$), 53 (17%) met the cut off for mild anxiety, 42 (13%) for moderate anxiety and 68 (22%) for severe anxiety.

5. Participants

Two hundred ten (67%) participants were female and 104 (33%) were male; the minimum age was 18 and the maximum age was 80 (*M* = 49.15, *SD* = 14.39). The majority of the participants were: married/cohabitating (*n* = 188, 60%); white (*n* = 214, 68%); and reported a 2009 annual income of less than $40,000 (*n* = 185, 64%). Forty participants (13%) reported occupations affected by the oil spill, including hospitality and tourism; seafood related industries; fishing; and oil/drilling support. The majority of participants were from parishes (counties) legally defined as most exposed to the DWH Gulf Oil Spill (*n* = 270, 86%), which include Lafourche, St. Bernard, Plaquemines, Terrebonne, Jefferson, and Orleans [20]. Seventy-nine or 25% applied for financial assistance following the oil spill. Participants were asked if they were a litigant in the BP lawsuit; 34 (11%) replied yes and 275 (89%) said no.

6. Statistical Analysis

To answer the primary hypothesis—sample equivalence on somatic complaints, posttraumatic stress, serious mental illness, anxiety and depression—two one-sided test (TOST) procedures were used with confidence intervals based on the Cohen's *d* *t*-test effect sizes to determine the margin of equivalence [31,32]. TOST procedures utilize traditional hypothesis difference testing (paired sample *t*-test), but extend the application to equivalence testing by asking whether the non-significant difference is small enough to determine that the samples are indeed similar [31]. United States Food and Drug equivalence determination was used and based on whether the mean difference lies within the confidence interval of equivalence [32]. To answer the secondary hypothesis—continued symptomatology would be associated with greater perceived disruption from the DWH Gulf Oil Spill—ordinary least squares regression was used. Regression was used to explore additional factors that may also contribute to continued levels of anxiety, such as demographics, Hurricane Katrina losses, and additional oil spill variables.

7. Results

The first step in assessing the hypotheses—sample equivalence on posttraumatic stress, serious mental illness, anxiety and depression—was to conduct five paired sample t-tests. Results are presented in Table 1, where results failed to reveal a significant difference on posttraumatic stress, serious mental illness, anxiety and depression. Next confidence intervals of equivalence were calculated based on Cohen's *d* to assess if the non-significant difference is small enough (see Table 1). Results revealed that the mean difference for anxiety lay within the margin of equivalence. While there was no statistical difference among posttraumatic stress, serious mental illness, and depression, the margin of equivalence did not include the mean difference between Time 1 and Time 2.

Given partial support of the primary hypothesis with no change in anxiety symptoms, regression analyses were used to explore which factors (being married or cohabiting, pre-oil spill income, oil/Gulf dependent occupation, litigation status, oil spill concerns, oil spill disruption, post spill funding requests, Hurricane Katrina losses) predict continued levels of anxiety. Preliminary analyses revealed that gender, race (white *vs.* other), parish (most impacted *vs.* other) r^{pb}-values (314) −0.01 to 0.03, *p*-values 0.63 to 0.86, and age, *r* (314) −0.01, *p* = 0.90, were not associated with anxiety thus these were not included in the regression. The enter method was used and with all variables accounted for 37% (adjusted R^2 = 0.355) of the variance in anxiety, F (8, 305) = 22.51, *p* < 0.001. Beta coefficients are presented in Table 2, where marital status, applied for financial assistance following spill, Hurricane Katrina losses, and oil spill disruption individually predicted anxiety. Results suggest that as individuals tend to be married or cohabitate, anxiety scores decrease by 0.11. For individuals that applied for financial assistance following the oil spill, anxiety scores decrease by 0.12. For individuals reporting a pre-oil spill income under $40,000, anxiety scores increase by 0.16. As Hurricane Katrina losses increase by 1, anxiety scores increase by 0.19 and as oil spill disruption increases by 1, anxiety scores increase by 0.42.

Table 1. Paired Sample Statistics and TOST Procedures for Mental Health Symptoms.

Mental Health	Time 1		Time 2		M^Δ (change)	95% CI Difference		*t*	*df*	*p*	*d*	90% CI Equivalence	
	M	*SD*	*M*	*SD*		L	U					L	U
Anxiety	8.0	7.2	8.0	7.1	−0.05	−1.01	0.90	−0.11	171	0.91	0.01	−0.14	0.16
Serious Mental Illness	6.8	6.6	6.2	6.6	0.65	−0.04	1.34	1.84	301	0.07	0.11	−0.01	0.22
Depression	9.7	9.3	10.3	9.5	−0.63	−1.98	0.73	−0.91	170	0.36	0.07	−0.08	0.22
Posttraumatic Stress	35.0	16.6	34.0	18.2	0.94	−0.95	2.83	0.98	292	0.33	0.06	−0.06	0.17

Table 2. Beta Coefficients Predicting Anxiety.

	B	SE	β	T	*p*	95% CI	
						Lower	Upper
Married or cohabitating	−1.53	0.69	−0.11	−2.21	0.028	−2.89	−0.17
Oil/Gulf dependent occupation	1.36	1.13	0.06	1.20	0.231	−0.87	3.58
Litigant	1.35	1.14	0.06	1.18	0.237	−0.89	3.59
Oil spill concerns	0.10	0.16	0.03	0.59	0.559	−0.23	0.42
Oil spill disruption	0.73	0.10	0.42	7.16	0.000	0.53	0.93
Hurricane Katrina Losses	0.59	0.15	0.19	3.89	0.000	0.29	0.90
Income above 40,000	−2.41	0.72	−0.16	−3.36	0.001	−3.82	−1.00
Post spill financial assistance	−2.02	0.89	−0.12	−2.26	0.025	−3.78	−0.26

8. Discussion

During the first 18 months following the Deepwater Horizon (DWH) Gulf Oil Spill, residents of Southeastern Louisiana reported increased symptoms of anxiety, depression, and posttraumatic stress [19,33]. The current study resampled individuals from the initial responders and results failed to reveal significant changes in anxiety, depression, serious mental illness, and posttraumatic stress two years post spill. Analyses further revealed that immediate anxiety symptoms were statistically equivalent to the elevated anxiety symptoms over two years following the disaster. While posttraumatic stress, serious mental illness, and depression did not statistically decrease, they were not statistically equivalent either. An explanation for anxiety statistically remaining at the same rates over two years post disaster may be contributed to the nature of the disaster. The role of uncertainty and unknown outcomes in a human caused disaster leads to anxiety on how, when, and if recovery will happen [1,22]. These findings suggest that the longer-term recovery trajectories for the DWH Gulf Oil Spill do not fall within the more traditional 18-month disaster recovery timeline [15,16,34].

Variables associated with continued symptoms of anxiety included marital status, application for financial assistance following the spill, Hurricane Katrina losses, and oil spill disruption. As with the initial study of immediate mental health symptoms following the spill [19], oil spill disruption was the most significant contributor to increased symptomotology, and accounted for the largest proportion of variance in anxiety symptoms. Interestingly an indirect association was revealed, where individuals that applied for financial assistance following the oil spill reported fewer symptoms of anxiety. This finding may support reports that the application process was overly complicated and was unattainable for the business practices of self-employed individuals in the fishing industries [23,24]. Contrary to the Exxon Valdez findings of Picou, Marshall and Gill [35], the low association between anxiety and litigation was no longer significant when accounting for the other variables. However, similar to their study, socioeconomic status predicted anxiety [35]; for individuals with incomes below $40,000 reporting more symptoms.

With rates of serious mental illness at 20%, depression at 35%, posttraumatic stress at 21% and moderate to severe anxiety at 35%, the rates of longer-term mental health symptoms continue to be elevated well above national norms of 6% for serious mental illness [36], 10% depression [37], 3% posttraumatic stress [38], and 18% for anxiety [39]. Mental health services are currently provided on a limited basis through the Gulf Region Health Outreach Program as part of the Deepwater Horizon Medical Benefits Class Action Settlement, which was approved by the U.S. District Court in New Orleans on 11 January 2013 and became effective on 12 February 2014. Four institutions from each of the four most impacted states collaborate to carry out the Mental and Behavioral Health Capacity Project (MBHCP), including the University of South Alabama, University of West Florida, Louisiana State University Health Sciences Center, and the University of Sothern Mississippi. A recent report on the Louisiana component of the project, supports the findings from the current study, and indicates a continued need for mental and behavioral health treatment [40].

The primary limitations with this study, consistent with disaster research [41], are the lack of pre-disaster data and reliance on self-report measures. While purposive sampling allowed for better representation of those directly affected by the spill, it does limit the generalizability to the larger populations. Other limitations include the relatively low response rate and the range of 17 months between Time 1 and Time 2. Analyses comparing respondents (33%) *versus* non-respondents (28%) on anxiety cut-off scores failed to reveal a significant group difference $\chi^2 = 2.3$, $p = 0.12$ or an association among time and anxiety ($r = -0.06$). Nonetheless, the low response rate and time between studies may have impacted findings in unknown ways. Another reason for lack of response may have been the ongoing litigation and fear that participation may be used against them in the settlement procedures. This limitation may have contributed to the lack of association among litigation and anxiety. Finally, the lack of litigation association may also suggest a limitation with timing due to ongoing legal action possibly influencing respondents to be hesitant to acknowledge their involvement. Continued longitudinal community surveys would help to better understand the overall recovery trajectory for individuals affected by the DWH Gulf Oil Spill. Further in-depth investigation of individuals that were most disrupted would provide more information to inform methods of how to address elevated symptoms.

9. Conclusions

This study supports many of the lessons learned from the Exxon Valdez spill, [3,8–10] suggesting that the indirect effects of the DWH Gulf Oil Spill are long term and recovery is slow. With mental health symptoms of anxiety, depression, PTSD and serious mental illness elevated above national rates, the need for continued mental health services is evident. Based on the above research mental health services should be targeted toward individuals with high levels of disruption and anxiety. In addition this study highlights the need for policy discussions around disaster recovery timelines and established norms [13].

Author Contributions: H.J.O., J.D.O. and A.S. conceived of the study. H.J.O., J.D.O. and T.C.H. designed the sampling and analysis plan. T.C.H compiled, summarized, interpreted the results and drafted the manuscript. All authors read, contributed to, and approved the manuscript.

Conflicts of Interest: The authors declare no conflict of interest.

References

1. Goldstein, B.D.; Osofsky, H.J.; Lichtveld, M.Y. The Gulf oil spill. *N. Engl. J. Med.* **2011**, *364*, 1334–1348. [PubMed]
2. Palinkas, L.A.; Petterson, J.S.; Russell, J.C.; Downs, M.A. Community patterns of psychiatric disorders after the Exxon Valdez oil spill. *Am. J. Psychiatry* **1993**, *150*, 1517–1524. [PubMed]
3. Palinkas, L.A.; Russell, J.C.; Downs, M.A.; Petterson, J.S. Ethnic difference in stress: Coping and depressive symptoms after the Exxon Valdez oil spill. *J. Nerv. Mental Disorder* **1992**, *180*, 287–295. [CrossRef]
4. Lyons, R.A.; Temple, J.M.; Evans, D.; Fone, D.L.; Palmer, S.R. Acute health effects of the Sea Empress oil spill. *J. Epidemiol. Commun. Health* **1999**, *53*, 306–310. [CrossRef]

J. Mar. Sci. Eng. **2015**, *3*, 1260–1271

5. Gallacher, J.; Brostering, K.; Palmer, S.; Fone, D.; Lyon, R.S. Symptomatology attributable to psychological exposure to a chemical incident: A natural experiment. *J. Epidemiol. Commun. Health* **2007**, *61*, 506–512. [CrossRef] [PubMed]

6. Carrasco, J.M.; Perez-Gomez, B.; Garcia-Mendizabal, M.J. Health-related quality of life and mental health in the medium-term aftermath of the Prestige oil spill in Galiza (Spain): A cross-sectional study. *BMC Public Health* **2007**, *7*, 245–252. [CrossRef] [PubMed]

7. Sabucedo, J.M.; Arce, C.; Senra, C.; Seoane, G.; Vazquez, I. Symptomatic profile and health-related quality of life of persons affected by the Prestige catastrophe. *Disasters* **2009**, *34*, 809–820. [CrossRef] [PubMed]

8. Palinkas, L.A.; Petterson, J.S.; Russell, J.C.; Downs, M.A. Ethnic differences in symptoms of posttraumatic stress after the Exxon Valdez oil spill. *Prehospital Disaster Med.* **2004**, *19*, 102–112. [PubMed]

9. Picou, S.; Formichella, C.; Marshall, B.; Arata, C. Community Impacts of the Exxon Valdez Oil Spill: A synthesis and elaboration of social science research. In *Synthesis: Three Decades of Research on Socioeconomic Effects Related to Offshore Petroleum Development in Coastal Alaska*; Stephen R. Braund & Associates: Anchorage, AK, USA, 2009; pp. 279–310.

10. Picou, S.; Arata, C. Chronic Impacts of the Exxon Valdez Oil Spill: Resource Loss and Commercial Fishers. In *Coping with Technological Disasters*; Prince William Sound Regional Citizens' Advisory Council: Anchorage, AK, USA, 1997; pp. J2–J43.

11. National Commission on the BP Deepwater Horizon Oil Spill and Offshore Drilling. *Deep Water: The Gulf Oil Disaster and the Future of Offshore Drilling*; Report to the President of the USA: New Orleans, LA, USA, 2011.

12. Gill, D.A.; Picou, J.S. Technological disaster and chronic community stress. *Soc. Nat. Resour.* **1998**, *11*, 795–815. [CrossRef]

13. McFarlane, A.C.; Williams, R. Mental Health Services Required after Disasters: Learning from the Lasting Effects of Disasters. *Depress Res. Treat.* **2012**, *2012*. [CrossRef]

14. Danya Institute. Disaster Mental Health Responder Certification Training at the DC Department of Behavioral Health. Available online: http://www.danyainstitute.org/2014/02/disaster-mental-health-responder-certification-training-at-the-dc-department-of-behavioral-health/ (accessed on 19 October 2015).

15. Substance Abuse and Mental Health Services Administration (SAMHSA). *Field Manual for Mental Health and Human Service Workers in Major Disasters*; ERIC: Washington, DC, USA, 2000.

16. Federal Emergency Management Agency. Deadlines & Timelines. Available online: http://www.fema.gov/public-assistance-local-state-tribal-and-non-profit/deadlines-timelines (accessed on 19 October 2015).

17. Kessler, R.C.; Sonnega, A.; Bromet, E.; Hughes, M.; Nelson, C.B. Posttraumatic stress disorder in the National Comorbidity Survey. *Arch. General Psychiatry* **1995**, *52*, 1048–1060. [CrossRef]

18. Kessler, R.C.; Galea, S.; Gruber, M.J.; Sampson, N.A.; Ursano, R.J.; Wessely, S. Trends in mental illness and suicidality after Hurricane Katrina. *Mol. Psychiatry* **2008**, *13*, 374–384. [CrossRef] [PubMed]

19. Osofsky, H.J.; Osofsky, J.D.; Hansel, T.C. Deepwater Horizon Oil Spill: Mental health effects on residents in heavily affected areas. *Disaster Med. Public Health Prep.* **2011**, *5*, 280–286. [CrossRef] [PubMed]

20. Abramson, D.M.; Redlener, I.E.; Stehling-Ariza, T.; Sury, J.; Banister, A.N.; Park, Y.S. *Impact on Children and Families of the Deepwater Horizon Oil Spill: Preliminary Findings of the Costal Population Impact Study*; Report for National Center for Disaster Preparedness: New York, NY, USA, 2010.

21. Witters, D. *Gulf Coast Residents Worse of Emotionally after BP Oil Spill*; Gallup: Washington, DC, USA, 2010.

22. Morris, J.G.; Grattan, L.M.; Mayer, B.M.; Blackburn, J.K. Psychological responses and resilience of people and communities impacted by the Deepwater Horizon oil spill. *Trans. Am. Clin. Climatol.* **2013**, *124*, 191–201.

23. Buttke, D.; Vagi, S.; Bayleyegn, T.; Sircar, K.; Strine, T.; Morrison, M.; Allen, M.; Wolkin, A. Mental health needs assessment after the gulf coast oil spill—Alabama and Mississippi, 2010. *Prehosp. Disaster Med.* **2012**, *27*, 401–408. [CrossRef] [PubMed]

24. Grattan, L.M.; Roberts, S.; Mahan, W.T.; McLaughlin, P.K.; Otwell, W.S.; Morris, J.G. Early psychological impacts of the Deepwater Horizon oil spill on Florida and Alabama communities. *Environ. Health Perspect.* **2011**, *119*, 838–843. [CrossRef] [PubMed]

25. Substance Abuse and Mental Health Services Administration and Centers for Disease Control and Prevention. *Behavioral Health in the Gulf Coast Region Following the Deepwater Horizon Oil Spill*; HHS Publication No. (SMA) 13-4737; Rockville, M.D., Ed.; Substance Abuse and Mental Health Services Administration and Centers for Disease Control and Prevention: Atlanta, GA, USA, 2013.

26. Sheehan, D.V.; Harnett-Sheehan, K.; Raj, B.A. The measurement of disability. *Int. Clin. Psychopharmacol.* **1996**, *11*, 89–95. [CrossRef] [PubMed]

27. Kessler, R.C.; Andrews, G.; Colpe, L.J.; Hiripi, E.; Mroczek, D.K.; Normand, S.-L.T.; Walters, E.E.; Zaslavsky, A. Short screening scales to monitor population prevalence and trends in nonspecific psychological distress. *Psychol. Med.* **2002**, *32*, 959–976. [CrossRef] [PubMed]

28. Weathers, F.W.; Litz, B.T.; Herman, D.S.; Huska, J.A.; Keane, T.M. The PTSD Checklist (PCL): Reliability; Validity; and Diagnostic Utility. In *Annual Convention of the International Society for Traumatic Stress Studies*; International Society for Traumatic Stress Studies: San Antonio, TX, USA, 1993.

29. Radloff, L.S.; Locke, B.Z. The community mental health assessment survey and the CES-D Scale. In *Community Surveys of Psychiatric Disorders*; Weissman, M.M., Myers, J.K., Ross, C.E., Eds.; Rutgers University Press: New Brunswick, NJ, USA, 1986; pp. 177–189.

30. Spitzer, R.L.; Kroenke, K.; Williams, J.B.W.; Lowe, B. A brief measure for assessing generalized anxiety disorder. *Arch. Inern. Med.* **2006**, *166*, 1092–1097. [CrossRef] [PubMed]

31. Wuensch, K.L. *Confidence Intervals; Pooled and Separate Variances T*; Department of Psychology, East Carolina University: Greenville, NC, USA, 2010.

32. Lachenbruch, P.A. *Equivalence Testing*; United States Food and Drug Administration: Silver Spring, MD, USA, 2001.

33. Osofsky, H.J.; Hansel, T.C.; Osofsky, J.D.; Speier, A. Factors Contributing to Mental and Physical Health Care in a Disaster-Prone Environment. *Behav. Med.* **2015**, *31*, 131–137. [CrossRef] [PubMed]

34. Center for Disease Control and Prevention. Disaster Mental Health Primer: Key Principles, Issues and Questions. 2012. Available online: http://emergency.cdc.gov/mentalhealth/primer.asp (accessed on 19 October 2015).

35. Picou, J.S.; Marshall, B.K.; Gill, D.A. Disaster; Litigation; and the Corrosive Community. *Soc. Forces* **2004**, *82*, 1493–1522. [CrossRef]

36. Kessler, R.C.; Chiu, W.T.; Demler, O.; Walters, E.E. Prevalence; severity; and comorbidity of twelve-month DSM-IV disorders in the National Comorbidity Survey Replication (NCS-R). *Arch. General Psychiatry* **2005**, *62*, 617–627. [CrossRef] [PubMed]

37. Center for Disease Control and Prevention. An Estimated 1 in 10 U.S. Adults Report Depression. Available online: http://www.cdc.gov/features/dsdepression/ (accessed on 19 October 2015).

38. National Institute of Mental Health. The Numbers Count: Mental Disorders in America 2010. Available online: http://www.nimh.nih.gov/health/publications/the-numbers-count-mental-disorders-in-america/index.shtml (accessed on 19 October 2015).

39. Anxiety and Depression Association of America. Facts & Statistics. Available online: http://www.adaa.org/about-adaa/press-room/facts-statistics (accessed on 19 October 2015).

40. Osofsky, H.J.; Osofsky, J.D.; Wells, J.H.; Weems, C. Integrated care: Meeting mental health needs after the Gulf oil spill. *Psychiatr. Serv.* **2014**, *65*, 280–283. [CrossRef] [PubMed]

41. Masten, A.S.; Osofsky, J.D. Disasters and their impact on child development: Introduction to the special section. *Child Dev.* **2010**, *81*, 1029–1039. [CrossRef] [PubMed]

Journal of
Marine Science and Engineering

MDPI

Article

Bitumen on Water: Charred Hay as a PFD (Petroleum Flotation Device)

Nusrat Jahan [1], Jason Fawcett [1], Thomas L. King [2], Alexander M. McPherson [1], Katherine N. Robertson [1], Ulrike Werner-Zwanziger [3] and Jason A. C. Clyburne [1,*]

[1]   Atlantic Centre for Green Chemistry, Departments of Chemistry and Environmental Science, Saint Mary's University, Halifax, NS B3H 3C3, Canada; drnusrat2001@hotmail.com (N.J.); jafawcett@gmail.com (J.F.); mcpherson.alexm@gmail.com (A.M.M.); Katherine.Robertson@smu.ca (K.N.R.)
[2]   Centre for Offshore Oil, Gas and Energy Research, Bedford Institute of Oceanography, Dartmouth, NS B2Y 4A2, Canada; Tom.King@dfo-mpo.gc.ca
[3]   Department of Chemistry and Institute for Research in Materials, Dalhousie University, Halifax, NS B3H 4J3, Canada; ulrike.wernerzwanziger@gmail.com
*   Author to whom correspondence should be addressed; Jason.Clyburne@smu.ca; Tel.: +1-902-420-5827.

Academic Editors: Merv Fingas and Tony Clare
Received: 6 August 2015; Accepted: 12 October 2015; Published: 20 October 2015

Abstract: Global demand for petroleum keeps increasing while traditional supplies decline. One alternative to the use of conventional crude oils is the utilization of Canadian bitumen. Raw bitumen is a dense, viscous, semi-liquid that is diluted with lighter crude oil to permit its transport through pipelines to terminals where it can then be shipped to global markets. When spilled, it naturally weathers to its original form and becomes dense enough to sink in aquatic systems. This severely limits oil spill recovery and remediation options. Here we report on the application of charred hay as a method for modifying the surface behavior of bitumen in aquatic environments. Waste or surplus hay is abundant in North America. Its surface can easily be modified through charring and/or chemical treatment. We have characterized the modified and charred hay using solid-state NMR, contact angle measurements and infrared spectroscopy. Tests of these materials to treat spilled bitumen in model aquatic systems have been undertaken. Our results indicate that bitumen spills on water will retain their buoyancy for longer periods after treatment with charred hay, or charred hay coated with calcium oxide, improving recovery options.

Keywords: petroleum; bitumen; dilbit; crude oil; asphaltene; remediation; spill; recovery; hay; flotation

1. Introduction

The development of the Canadian oil sands in northern Alberta has become a significant contributor to the Canadian economy. It could also become a significant contributor of oil to the world economy. Bituminous sands in Canada have been assessed to hold approximately 43% of the total global bitumen deposits, which represents approximately 26.9 billion m^3 or about 169.3 billion barrels of crude bitumen [1]. The locations of these Canadian bitumen deposits are far from both ocean and refinery access. Using current techniques crude bitumen can be refined to approximately 20% by weight of petroleum coke. This is of little value while landlocked but it could be valuable, even though high in sulfur, in "leading edge" environmental applications such as a source of activated carbon to reduce the toxins content in oil sands tailings [2]. In order to produce and transport bitumen profitably, ocean access for shipment is essential. Possible transportation routes might include: (1) railway to refinery or ocean port; (2) pipeline to railway or refinery or ocean port; or (3) truck to railway or refinery or ocean port.

The nature of these transportation methods, and the frequency and volume of product being transported, increases the risk of accidental spills or pipeline leaks. A review of the scientific literature indicates that there are limited options available to treat diluted bitumen (dilbit) spills. An article published by the Royal Society of Canada in 2010 assessed the environmental and health impacts of Canada's oil sands industry. This report outlines and summarizes a number of areas of concern regarding the environmental impacts of the oil sands industry, and suggests some of the necessary reclamation and monitoring practices necessary to mitigate these impacts [3]. A later 2012 publication in the journal, Environmental Science and Technology, discussed a number of shortcomings and oversights in the 2010 assessment [4]. However, both reports neglect to mention the need for new oil spill treatment technologies, given that bitumen and dilbit, under the correct circumstances, will sink. Conventional technologies, such as dispersants, will be rendered less effective in the event of a major dilbit spill either on site (e.g., land based) or during transport (e.g., entering aquatic systems).

The physical properties and composition of crude bitumen make it a very challenging material to manipulate. Bitumen is a heavy, viscous, semi-solid form of petroleum and is composed of a complex mixture of materials. At 15 °C, the complex viscosity of Athabasca bitumen has been reported to be 1.75×10^7 mPa·s [5]. The dynamic viscosity is reported to range from 1.9×10^4 to greater than 7.0×10^5 mPa·s at the same temperature [6], compared to conventional heavy crude, such as heavy fuel oil, HFO 6303, which has a reported viscosity of 2.28×10^4 mPa·s under the same conditions [6]. The same reference [6] reports the density of Athabasca bitumen as being 1.006 to 1.016 g cm^{-1}. A 2011 report in the Journal of Chemical and Engineering Data [5] details "Saturates, Aromatics, Resins and Asphaltenes" (SARA) analyses and Mass Fraction results which have been used to characterize the composition of Athabasca bitumen. These findings are outlined in Table 1 and notably include an asphaltene component of 18.6% ± 1.86% by weight in the bitumen studied.

Table 1. Composition analyses of Athabasca bitumen (adapted from Bazyleva *et al.* [5], with permission from © 2011 American Chemical Society).

Elemental Composition	Weight %
Carbon	83.2 ± 0.9
Hydrogen	9.7 ± 0.4
Nitrogen	0.4 ± 0.2
Sulphur	5.3 ± 0.2
Oxygen	1.7 ± 0.3
SARA Analysis	**Weight %**
Saturates	16.1 ± 2.1
Aromatics	48.5 ± 2.3
Resins	16.8 ± 1.2
Asphaltene (C$_5$)	18.6 ± 1.8

A series of publications spanning the years 2010 to 2012 by Murray R. Gray *et al.*, have tackled the arduous challenge of characterizing the structures of various bitumen fractions [7–9]. The most significant component of bitumen, the one that differentiates it from conventional crude oil, is the abundant asphaltene fraction. Asphaltenes are the heaviest fraction of crude bitumen, and consist mostly of polycyclic-aromatic rings complexed with metals including nickel and vanadium.

Asphaltenes are problematic for bitumen processing for a number of reasons, arising mainly by their tendency towards aggregation. Aggregation occurs because of various acid-base interactions, hydrogen bonding and the formation of metal-containing coordination complexes. This aggregation results in the drastically higher viscosity observed for crude bitumen as compared to crude oil. This, in turn, gives rise to the observed difficulties in pumping and processing bitumen. In the case of an ocean or fresh water-based bitumen spill aggregation will more than likely result in the clumping and sinking of the spilled materials. Understanding the aggregation behavior of asphaltenes in bituminous oils is essential to developing methods and materials for spill treatment/recovery.

A recent report from the Federal Government of Canada assesses the spill behavior and fate of two diluted bitumen (dilbit) samples under different weathering conditions. The dilbit products selected were those most frequently transported in Canada. Preliminary laboratory investigations showed that the dilbit products remained buoyant under natural ocean-simulated weathering conditions (0–15 °C) except when mixed with fine to moderately sized sediments [10]. One gap in this investigation was that only two samples of dilbit (e.g., Cold Lake Blend and Access Western Blend) were tested, and they were not compared to a base sample of crude bitumen. Furthermore, the products were studied only in sea water conditions. It must be remembered that there is also significant risk of spills occurring in fresh or brackish waters.

To extend the initial results to such waters, King *et al.*, (2014) have investigated dilbit weathering, through meso-scale (e.g., wave tank) studies, under natural conditions. One of the same dilbit products (e.g., Access Western Blend) was shown to weather enough, without interaction with sediments, such that its density exceeded that of fresh and brackish waters [11,12]. The authors concluded that this product would initially float on aquatic systems, but that after 6 days of natural attenuation, the product would sink in aquatic systems. A very recent paper by Stevens *et al.*, offers proof that oil weathering can result in its sinking [13]. The authors have developed an evaporation/sinking (EVAPOSINK) model that can be used to predict such behavior.

The potential for diluted bitumen products to sink when spilled is problematic from both environmental and industrial perspectives. Sunken oil is more difficult to find and track, and there are no known spill countermeasures to treat submerged dilbit. Preliminary findings have shown dispersants to be ineffective in the treatment of a diluted bitumen spill [10]. Submerged oil could potentially cause significant and persistent loss of potable water, ecosystems (e.g., rivers and lakes, marine systems, *etc.*) and aquatic life. Further investigation into the spill behavior of crude bitumen in aquatic systems is essential for the development of a cheap and effective countermeasure for spill impact mitigation and recovery. There is a definite need to identify a material capable of reducing the bulk density of the bitumen to keep it floating on the aquatic surface for as long as possible. This would prolong the window of opportunity available during the flotation phase to treat the spill by either mechanical means, such as booming or skimming the surface, or through *in situ* combustion.

Our preliminary investigations led us to hay, a cheap and abundant material with a large surface area. We felt that it might be a suitable material to adsorb bitumen and act as a flotation device. Attempts were made to modify the surface of the green hay so that it would also act as a natural dispersant. The hay was first immersed and coated with the organic-based surfactant, "Zep", a limonene-based household degreasing product. This surface modification was unsuccessful; the surfactant did not result in a modification of the surface properties of the hay. When this preliminary treatment failed, charring the hay and/or coating it with calcium oxide were investigated as means of surface modification. It was anticipated that charring the hay surface would render it more hydrophobic by removing surface OH groups and exposing the carbon backbone, while addition of CaO could possibly generate an *in situ* surfactant, improving the dispersant properties of the system. The results of the investigation are reported herein.

2. Experimental Section

2.1. Chemicals, Oils and Oil Spill Treating Agents

Athabasca bitumen was provided by the Centre for Oil Sands Innovation, Edmonton, Alberta and was used as received. Timothy hay (*Phleum pretense*) was purchased at Walmart, as supplied by Pestell Pet Products of Ontario, Canada. The composition of the hay is listed as follows: crude protein (min. 7.5%), crude fat (min. 2.0%), crude fibre (max. 35%), moisture (max. 12.0%), and calcium (min. 0.25%–max. 0.60%). "Instant Ocean" Sea Salt is distributed by United Pet Group Inc.

of Cincinnati, OH, USA. It was prepared as directed on the packaging. "Zep" Heavy Duty Citrus Degreaser, with the active ingredients, d-limonene and monoethanolamine, was obtained from the Home Depot (Zep Superior Solutions, Atlanta, GA, USA). Reagent grade nitric acid, ACS reagent grade dichloromethane ≥99.5%, PCR reagent grade chloroform ≥99% and reagent grade calcium oxide and potassium bromide were purchased from Sigma-Aldrich Canada (Oakville, ON, Canada) and used as obtained.

2.2. Experimental Design

Simulated bitumen slicks were prepared in 250 mL glass beakers. A measured volume of 100 mL of either deionized water or Instant Ocean solution was added to each beaker. Bitumen slicks were generated by applying a known mass (1.8 g) of crude bitumen to the surface for both fresh water and Instant Ocean (artificially created salt water) samples. Samples were left at room temperature (23 °C) and were stirred for 2–3 min every 12 h. Samples were also periodically photographed to record bitumen aggregation and subsequent sinking of the product over time. At the end of the observation period, samples of the experimental solutions were collected. These were analyzed for trace metals, total petroleum hydrocarbon content and density.

Hay samples were cut into lengths of approximately 1.0 cm, sufficiently short to fit into the experimental beakers. The cut hay (all from a single source) was mixed to randomize its distribution before use, but no other attempts were made to homogenize the hay in the samples and replicate measurements were not performed for the charring process itself. The surface properties of the straw were then altered as follows: (1) the hay was charred to remove hydrogen and oxygen from its surface; and/or (2) the hay was coated with calcium oxide for, potentially, *in situ* surfactant formation. Addition of bitumen to the CaO-treated hay could possibly result in the deprotonation of the carboxylic acids, which would generate an *in situ* surfactant.

(1) The charred hay was prepared by placing the clippings in a sealed Schlenk flask and then placing the flask under vacuum. A propane torch was carefully applied to the bottom of the flask as it was mixed to endure uniform heating. Heating was performed at 10 min intervals, and the flask allowed to cool between heating cycles. Depending on the experiment, heating was continued for approximately 30 or 60 min total. During the hay charring process, a clean solvent trap was inserted into the Schlenk line and liquid nitrogen was used to condense the evolved gases. The condensate was washed from the trap using acetone, which was subsequently removed by evaporation. Preliminary experiments have been carried out to analyze the condensate for its principle components using gas chromatography coupled with mass spectrometry (GC/MS).

(2) Calcium oxide coated hay samples were prepared using the following procedure. A supersaturated solution of calcium carbonate (5 g) was prepared by adding just enough deionized water to make a paste. Then 2.5 g of uncharred hay clippings were mixed and coated with the paste and the mixture was left for 24 h at room temperature. A portion of the original mixture (the CaO-coated, uncharred hay sample) was then transferred to a Schlenk flask and charred under vacuum (see above) to produce the CaO-coated, charred hay samples. Heating was continued until the surface of the hay turned dark brown-black.

Buoyancy and bitumen adsorption of the charred hay samples (30 or 60 min) were evaluated by preparing sample slicks, containing approximately 2.0–2.3 g of bitumen in 100 mL of solution, as outlined above. The slicks were treated by adding 1.0 g of charred hay. Samples were shaken daily, and observed and photographed as outlined in the procedure above. A final set of buoyancy experiments examined the effectiveness of charred straw relative to CaO-coated charred straw. Instant Ocean solution (350 mL) was added to 125×65 mm^2 glass dishes to which were also added 3 g of bitumen and either 2 g of charred hay or 5 g of CaO-coated charred hay. The bitumen and straw were well mixed and then the dishes were placed on an orbital shaker operating at 65 rpm at room temperature. Once again, samples were observed and photographed periodically as outlined above.

2.3. Sample Analyses

2.3.1. Density

The densities of the deionized water and the Instant Ocean solution were measured by accurately determining the mass and volume of a specified quantity of each solution at room temperature. The density of Athabasca bitumen has been reported to be 1.006 to 1.016 g cm^{-1} [6].

2.3.2. Contact Angle Measurements

The differences in the *potential* strength of adsorption to the altered hay surfaces were evaluated via contact angle measurements of water droplets on flat surfaces of both the charred and the uncharred hay. Contact angle measurements were performed using a First Ten Angstroms (FTA) 135 Drop Shape Analyzer and FTA-32 Video software (Portsmouth, VA, USA).

2.3.3. Infrared Spectroscopy

Infrared spectra were recorded on a Bruker Vertex 70 Infrared Spectrometer (Bruker Optics Ltd., Milton, ON, Canada), with samples prepared as KBr pellets. Data processing was completed using OPUS 6.0 software (Bruker Optics Ltd., Milton, ON, Canada).

2.3.4. Nuclear Magnetic Resonance (NMR) Spectroscopy

The solid state ^{13}C cross polarization (CP)/magic angle spinning (MAS) NMR spectrum of raw hay was compared to those of two different samples of charred hay, one charred for 30 min and the other charred for 60 min. These NMR experiments were carried out in the NMR-3 Facility of Dalhousie University on a Bruker Avance DSX NMR spectrometer with a 9.4 Tesla magnet (400.24 MHz ^{1}H and 100.64 MHz ^{13}C Larmor frequencies) using a probe head for rotors of 4 mm diameter (Billerica, MA, USA). The parameters for the ^{13}C CP/MAS experiments with TPPM proton decoupling were optimized on glycine, whose carbonyl resonance also served as an external, secondary chemical shift standard at 176.06 ppm. For the final ^{13}C CP/MAS NMR spectra 1200 scans were acquired with 13.5 kHz sample spinning, 2.6 ms cross-polarization times and 3 s repetition times, as determined from the ^{1}H spin lattice relaxation times, T_1. Additional spectra, taken at 5.0 kHz sample spinning and also with a ^{13}C CP/MAS sequence followed by TOSS (TOtal Sideband Suppression), showed that there is no significant overlap between spinning sidebands and center bands.

2.3.5. Gas Chromatography with Flame Ionization Detection

Residual oil in the samples was analyzed using the method outlined by Cole *et al.* [12]. Briefly, the method is a modified version of EPA 3500C, where the sample container is used as the extraction vessel. Dichloromethane (DCM) was added to the sample bottle containing dispersed oil in solution. The sample was placed on a Wheaton R$_2$P roller (VWR International Ltd., Mississauga, ON, Canada) for 18 h. The roller had been modified to accommodate 3 inch diameter PVC pipe into each roller slot, so that sample containers of different sizes could be used. Once extraction was complete, the samples were removed and the DCM recovered. The recovered DCM was placed in a pre-weighed 15 mL centrifuge tube and the solvent volume reduced under a nitrogen evaporator to 1.0 mL. The extracts were analysed by gas chromatography using flame ionization detection. The original bitumen product was used to prepare calibration standards that were then used to generate a calibration curve from which oil concentrations in the extracts could be calculated. A mean percent recovery of 90.8 $\pm$ 4.6% was calculated from all oils spiked into water. The method detection limit was <0.5 mg/L. The method of extraction and analyses has been validated against the US EPA 3510C and provides better extraction efficiency for oils. The GC-FID method (EPA 8015B) is a standard US EPA method for analysing oils. The method has been published as supplementary material in an article in Environmental Engineering Science in 2015 [14].

2.3.6. Inductively Coupled Plasma Mass Spectrometry (ICP-MS)

All samples for ICP-MS analysis were freshly prepared. The required components for each sample, bitumen, green straw, charred straw, CaO-coated green straw, or CaO-coated charred straw, were placed into either deionized water or Instant Ocean solution. All straw-containing samples included 100 mL of solution (deionized water or Instant Ocean), 1 g of bitumen, and 0.5 g of straw (green or charred, with or without a CaO coating). The non-straw samples contained 100 mL of solution, 2 g of bitumen and in half of the samples added CaO (0.5 g). They were all left in the refrigerator for 48 h. The sample solutions were filtered through a 0.45 µm pore size (GHPP, Pall Gelman Acrodisc, purchased from Sigma-Aldrich Canada, Oakville, ON, Canada) syringe filter and acidified using 10% nitric acid to a pH of less than 2, prior to ICP-MS analyses for dissolved metals. Inductively Coupled Plasma Mass Spectrometry (ICP-MS) was performed at the Saint Mary's University Center for Environmental Analysis and Remediation (CEAR) on a VG PQ ExCell instrument (Thermo Elemental, Winsford, UK) by Patricia Granados.

3. Results and Discussion

3.1. Density

The densities of deionized water and the Instant Ocean solution were measured at 23 °C and determined to be 0.980 and 1.004 g/mL, respectively. Both are lower than a literature value reported for bitumen of 1.006 to 1.016 g/cm^3 [6].

3.2. Characterization of Charred versus Uncharred Hay

Preliminary experiments have been carried out on the condensate collected during the hay charring process. GC/MS analysis of the hay condensate showed that it contained many compounds, with three significant contributors being vanillin lactoside, 2,6-dimethoxyphenol (syringol) and 5-hydroxymethyl-2-furaldehyde. All of these are known components of combustion extracts [15].

3.2.1. Contact Angle Measurements

Contact angle measurements were made to characterize the impact of modification of the hay surface through the charring process on the bulk properties of the material. The expectation was that charring the hay would liberate hydrogen and oxygen from its surface, increasing C=C bond formation, and result in an overall increase in the hydrophobicity. Contact angle measurements for deionized water on charred and raw hay surfaces showed a significant increase in the contact angle after charring (e.g., from approximately 64°, Figure 1 (left) to approximately 126°, Figure 1 (right) and therefore, a definite change in the hydrophobic properties of the surfaces.

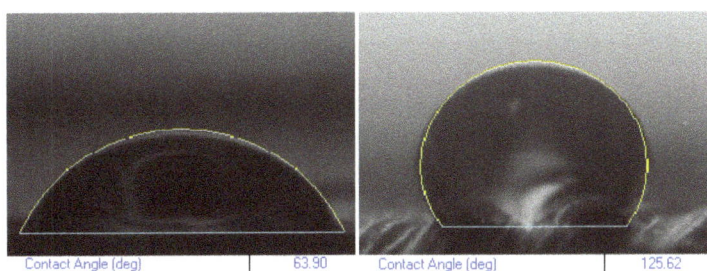

Figure 1. Contact angle measurements, before (**left**) and after (**right**) charring of the hay surface.

If the contact angle of water is less than 30°, a surface is designated as hydrophilic; wetting of the surface is favourable, and the water will spread over a large area. On a hydrophobic surface, water forms into distinct droplets. As the hydrophobicity increases, the contact angle of the droplets with the surface increases. Surfaces with contact angles greater than 90° are designated as hydrophobic [16]. The measured change in contact angle for our samples supports the conclusion that chemical modifications resulting from charring have produced a more hydrophobic surface. Variations in the observed contact angles were noted after subsequent measurements, however, samples consistently revealed larger contact angles, and thus increased hydrophobicity, after charring.

3.2.2. Infrared Spectroscopy

Infrared spectra were recorded for samples of both raw and charred hay. Transmittance in the 1600–1700 cm^{-1} (wavenumber) region was observed to be reduced after the hay was charred (boxed region in Figure 2). This has been attributed to the removal of absorbed water on charring, with a concomitant decrease in the observed OH bending signal. Charring should also expose the carbon backbone and thereby alter the hydrophobic properties of the hay surface. In this regard it is important to note the changes in the O–CH$_3$ methyl stretching region [17] at 2850–2815 cm^{-1} (* in Figure 2), and in the small peak at 809 cm^{-1} (C–C–O and C–O–C deformations) which disappears completely on charring. Other peaks also change in relative intensity, and all of this suggests that actual chemical modification has occurred. There is also a small possibility that these differences could have arisen from differences in the original hay samples themselves since the experiment was not performed on replicate samples. Although the spectra may appear only slightly different overall, the effect on the surface properties supports the evidence from the contact angle measurements that chemical modification of the surface has been achieved.

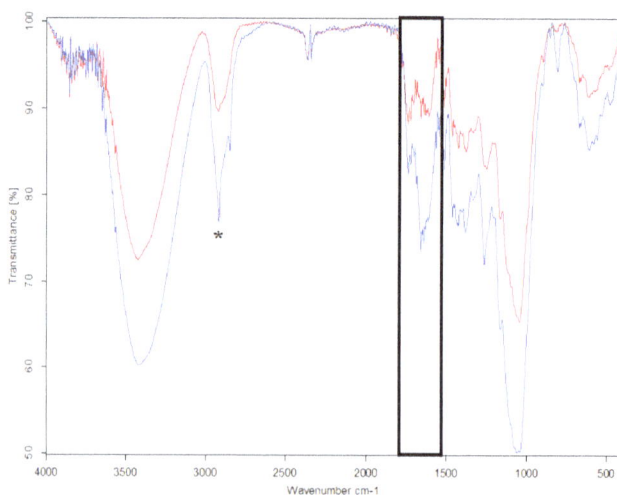

Figure 2. Overlay of infrared spectra comparing a raw hay sample (blue/bottom) to a charred hay sample (red/top). The boxed and starred areas show regions where the two spectra are distinctly different (see text).

3.2.3. ^{13}C CP/MAS Nuclear Magnetic Resonance Spectroscopy

All of the ^{13}C CP/MAS NMR spectra shown in Figure 2 exhibit typical cellulose signatures, with the alcoholic carbons between 50 and 90 ppm and the acetyl groups around 105 ppm. In addition, they show resonances for aliphatic groups between about 50 and 0 ppm. On the high chemical shift side, signals of carboxyl groups (around 173 ppm) and of unsaturated carbon groups, between about 110 and 155 ppm, of aromatic and possibly aliphatic origins are detected. In particular, the region between 140 and 155 ppm, corresponds to aromatic carbons bridging to other carbons or hetero-nuclei, such as oxygen. Comparison between the three samples shows that, relative to the largest peak, the intensities of the unsaturated (aromatic) region, the carboxyl groups and some aliphatic groups increase with increased charring time (indicated by *) (see Figure 3).

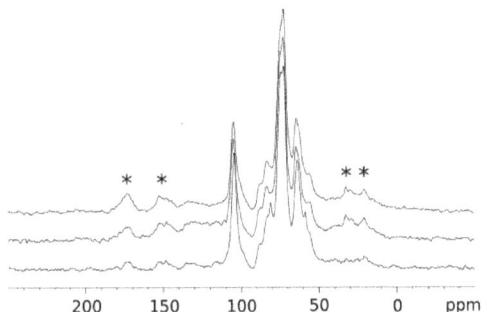

Figure 3. ^{13}C CP/MAS NMR spectra of raw hay (bottom) and charred hay samples (30 min middle, 60 min top) showing the relative intensity increases in the carboxyl, aryl and alkyl regions (isotropic center bands indicated by *) with charring time.

3.3. Application of Treated Hay to Bitumen Samples in Aqueous Environments

3.3.1. Flotation Analysis

A sample of bitumen in tap water with no additives was found to aggregate in the beaker and subsequently sink after a period of 4 days (96 h) at room temperature (23 °C). In the initial stages of the experiment, there was an even distribution of the bitumen on the surface of both the water solution and the Instant Ocean solution (Figure 4A,B, respectively). At 4 days, the bitumen sample in deionized water was observed to aggregate into a ball and sink to the bottom of the beaker on stirring (Figure 5A). As expected, the sample of bitumen in Instant Ocean solution remained afloat longer than the water sample because of the greater density of the Instant Ocean solution. As time passed the Instant Ocean solution became gradually more yellow in color and the bitumen slick increased in diameter (Figure 5B). It showed a tendency to aggregate and sink on stirring but it would remain suspended in the solution (never sinking to the bottom of the beaker) before rising again and dispersing across the surface of the solution. When the experiment was terminated after 121 h, the bitumen was still floating on the Instant Ocean solution.

Figure 4. Bitumen samples in deionized water (**A**) and Instant Ocean solution (**B**) at room temperature (23 °C) and time = 0 h.

Figure 5. Bitumen samples in deionized water (**A**) and Instant Ocean solution (**B**) at room temperature (23 °C) and time = 96 h.

In solution, bitumen did not interact to any great extent with green, uncharred hay. Instead, the hay just became waterlogged and sank while the bitumen floated on the surface of the solution. However, samples of bitumen treated with charred hay as a flotation additive were buoyant on the water surface for up to 186 h. The application of hay charred for 60 min to the bitumen slicks is shown in Figure 6 (at time = 0 h) and in Figure 7 (at time = 186 h, where both samples have sunk). Samples containing the hay charred for 60 min remained buoyant for equally long (Instant Ocean) or longer (water) than the samples containing hay charred for only 30 min. In fact, the bitumen and charred hay sample (30 min) sank in water after only 72 h. Comparing Figure 7 to Figure 6, dramatic color changes can be seen in the solutions (both water and Instant Ocean). These changes can be attributed to surface interactions between the charred hay and the hydrophobic fractions of the bitumen resulting in dissociation of some of the more polar fractions of the bitumen sample into the solutions over time. For the same reason, all of these samples are more highly colored than the samples of bitumen alone (Figure 5).

Figure 6. Bitumen samples with charred hay (60 min) added in deionized water (**left**) and Instant Ocean solution (**right**) at room temperature (23 °C) and time = 0 h.

Figure 7. Bitumen samples with charred hay (60 min) added in deionized water (**right**) and Instant Ocean solution (**left**) at room temperature (23 °C) and time = 186 h.

Samples of bitumen treated with charred hay (Figure 8) or CaO-coated charred hay (Figure 9), a flotation *and* dispersant additive, were buoyant for more than 408 h (17 days) at which point the experiment was terminated. In both samples, it is clear that the bitumen adsorbed to the surface of the hay undergoes dispersion. However, this dispersion appears to be greater when the charred straw has been treated with calcium oxide. It was also noted that the Instant Ocean solution did not yellow as much over time when CaO-coated hay was used as the dispersant. This may be the result of the increased dispersion or it may indicate a reduction in the fractional dissolution of the polar compounds with time for the coated sample. The latter idea is supported by the leached hydrocarbon analyses presented in the following section.

Figure 8. A bitumen sample in Instant Ocean treated with charred hay at room temperature (23 °C) and time = 288 h.

Figure 9. A bitumen sample in Instant ocean treated with CaO-coated charred hay sample at room temperature (23 °C) and time = 336 h.

3.3.2. Residual Oil Concentration

The solutions remaining in the beakers (both Instant Ocean and deionized water) at the end of the floatation experiments were analyzed for leached hydrocarbons. The liquid phase was separated from the hay and bitumen residues prior to measurement. As can be seen in Table 2, all samples containing bitumen showed leaching of hydrocarbons into solution, and this leaching was generally greater in Instant Ocean than in fresh water (deionized). Bitumen samples treated with raw hay in both Instant Ocean solution and deionized water showed slightly more leaching of petroleum hydrocarbons than any of the other treatments. However, this relative amount is statistically insignificant with p-values > 0.05 (2-factor ANOVA). Bitumen samples treated with only CaO showed the smallest quantity of leached petroleum hydrocarbons in both Instant Ocean solution and deionized water. We speculate that the strong base, CaO, either saponifies or deprotonates acidic species in the bitumen thus generating an *in situ* surfactant that better bonds the hydrocarbons to the charred straw.

Samples of bitumen treated with hay did show ppm levels of hydrocarbons (e.g., aliphatic, and parental and alkylated polycyclic aromatic hydrocarbons), entering the water phase at a slow rate over several days due to dispersion. The aromatics that enter the water column would contribute to toxicity in organisms exposed to the contaminated water. However, in an open marine environment, where there are no boundaries, these chemicals would spread over a greater spatial area and be exposed to natural dilution depending on sea states and environmental conditions. The natural dilution of these hydrocarbons would reduce their environmental impacts.

Table 2. Total Petroleum Hydrocarbons (TPH) in water samples collected from water column after treatment.

Sample	TPH (mg/L)	
	Instant Ocean Solution	Deionized Water
Standard (water)	<1.0	<1.0
Bitumen	47	28
Bitumen + raw hay	60	52
Bitumen + charred hay	54	21
Bitumen + CaO	18	18
Bitumen + CaO-coated raw hay	19	21
Bitumen + CaO-coated charred hay	52	29

3.3.3. Inductively Coupled Plasma—Mass Spectroscopy

ICP-MS experiments were performed under a variety of conditions in order to assess, if possible, any observable trends resulting from metal ion interaction (leached from the bitumen) with the biomaterial (hay). Reliable results were obtained from the fresh water samples only. Results of experiments performed in the Instant Ocean solutions were complicated due to interference from its high concentration of salts with the metal ions being studied. All samples were tested for V, Cr, Mn, Co, and Cu.

Analysis of the deionized water used in sample preparation showed no (or only trace levels of) V, Mn and Co. Levels of Cr measured 20 ppb and Cu 90 ppb. Analysis of the Instant Ocean solution showed that the product itself contained no Co or Cr, though Cr was present in the solution at the same level found in the water used to prepare it. Instant Ocean was also found to contain V 60 ppb, Mn 30 ppb and Cu 230 ppb (partially from the water).

Addition of bitumen to deionized water resulted in no change to the levels of V, Cr or Co measured. Mn was found to leach into the water, the concentration increasing from a trace (2 ppb) to 20 ppb. The opposite effect was observed for the levels of Cu in solution. Bitumen appears to absorb copper from the solution as the levels decreased from 90 to 30 ppb for Cu. The results when bitumen was added to Instant Ocean were similar. No change in the concentration was observed for V, Cr, Mn or Co. In the case of Mn there already was a concentration of 30 ppb in the solution which may have prevented more from leaching in from the bitumen. The bitumen in Instant Ocean also appeared to absorb some of the copper from solution.

The addition of straw (no CaO) to solutions of bitumen in deionized water had little observable effect on metal ion concentrations. V, Cr and Co levels were totally unaffected. The addition of green straw slightly increased the levels of Mn and Cu in the water, while charred straw had no effect. This may be because the hay itself contains both Cu and Mn which may leach more easily from the green straw. The addition of straw (no CaO) to bitumen in Instant Ocean did not affect the measured concentrations of Cr, Co or Cu. The level of V increased, while the level of Mn increased appreciably upon the addition of the straw, and in both cases the impact of the charred hay was greater than that of green hay.

The effect of a CaO-coating on the straw samples was assessed by comparing the green and charred straw numbers to the corresponding CaO-coated green and charred straw results. For the deionized water solutions the results, whether green or charred straw was used, were very consistent. The V, Cr and Co levels remained relatively constant while the Mn concentrations increased and the Cu concentrations decreased upon using straws coated with CaO. For the Instant Ocean solutions the results were a bit more scattered. However, overall the Cr, Co and Cu concentrations remained relatively constant while the V levels increased and the Mn levels decreased with the introduction of the CaO coating on the straws.

4. Conclusions

This work has shown that charred hay (and in particular CaO-coated charred hay) is an effective substrate for adsorption and flotation of bitumen. With further testing and fine-tuning it could become a valuable tool in the treatment of bitumen and dilbit spills (dilbit tests are ongoing) in aqueous environments. We have shown that its use should prolong the window of opportunity for skimming or *in situ* combustion of spilled oil by increasing the time the bitumen remains afloat. One clear advantage of charred hay is the strong interactions that form between its modified surface and the bitumen, limiting any washing out effect that might occur via wave action or weathering. Hay is also cheap, biodegradable and easily produced in bulk quantities. Additionally, charred hay demonstrates potential for prophylactic treatment on shorelines to prevent bitumen/dilbit from adhering to and contaminating shoreline materials such as sand, rocks and sensitive habitat.

Supplementary Materials: Figures S1 and S2—Contact Angle Measurements; Tables S1 and S2—Statistical Analysis of the Residual Oil Concentrations; Tables S3–S8—ICP-MS Data Summary Tables.

Acknowledgments: We thank the Natural Sciences and Engineering Research Council of Canada (through the Discovery Grants Program to JACC). JACC acknowledges generous support from the Canada Research Chairs Program, the Canadian Foundation for Innovation and the Nova Scotia Research and Innovation Trust Fund. We are grateful to NMR-3 (Dalhousie University) for NMR data acquisition and Patricia Granados (CEAR) for helpful discussions of the ICP-MS results. We would also like to acknowledge the Dalhousie University research group of J. Dahn for their assistance with contact angle measurements.

Author Contributions: Nusrat Jahan: Lab Experimentation; Jason Fawcett: Lab Experimentation; Thomas King: Residual Oil Analysis, Report Writing/Editing; Alex McPherson: Experiment Design, Lab Experimentation, Report Writing/Editing; Katherine N. Robertson: Data Analysis and Manuscript Preparation; Ulrike Werner-Zwanziger: NMR Data Acquisition, NMR Analysis and NMR Writing/Editing; Jason Clyburne: Funding, Intellectual Property, Experiment Design and Report Writing/Editing.

Conflicts of Interest: The authors declare no conflict of interest.

References

1. *ST98-2011: Alberta's Energy Reserves 2010 and Supply/Demand Outlook 2011–2020*; Energy Resources Conservation Board: Calgary, AB, Canada, 2011.
2. Zubot, W.; MacKinnon, M.D.; Chelme-Ayala, P.; Smith, D.W.; El-Din, G.M. Petroleum coke adsorption as a water management option for oil sands process-affected water. *Sci. Total Environ.* **2012**, *427–428*, 364–372. [CrossRef] [PubMed]
3. Gosselin, P.; Hrudey, S.E.; Naeth, M.A.; Plourde, A.; Therrien, R.; van Der Kraak, G.; Xu, Z. *The Royal Society of Canada Expert Panel: Environmental and Health Impacts of Canada's Oil Sands Industry*; The Royal Society of Canada: Ottawa, ON, Canada, 2011.
4. Timoney, K. Environmental and health impacts of Canada's bitumen industry: In search of answers. *Environ. Sci. Technol.* **2012**, *46*, 2496–2497. [CrossRef] [PubMed]
5. Bazyleva, A.; Fulem, M.; Becerra, M.; Zhao, B.; Shaw, J.M. Phase behavior of Athabasca bitumen. *J. Chem. Eng. Data* **2011**, *56*, 3242–3253. [CrossRef]
6. Environmental Science and Technology Division. Environment Canada Oil Spill Properties Database. Available online: http://www.etc-cte.ec.gc.ca/databases/OilProperties/ (accessed on 22 May 2014).
7. Borton, D., II; Pinkston, D.S.; Hurt, M.R.; Tan, X.; Azyat, K.; Scherer, A.; Tykwinski, R.; Gray, M.; Qian, K.; Kenttämaa, H.I. Molecular structures of asphaltenes based on the dissociation reactions of their ions in mass spectrometry. *Energy Fuels* **2010**, *24*, 5548–5559.

8. Gray, M.R.; Tykwinski, R.R.; Stryker, J.M.; Tan, X. Supramolecular assembly model for aggregation of petroleum asphaltenes. *Energy Fuels* **2011**, *25*, 3125–3134. [CrossRef]
9. Rueda-Velásquez, R.I.; Freund, H.; Qian, K.; Olmstead, W.N.; Gray, M.R. Characterization of asphaltene building blocks by cracking under favorable hydrogenation conditions. *Energy Fuels* **2013**, *27*, 1817–1829. [CrossRef]
10. *2013 Federal Government Technical Report: Properties, Composition and Marine Spill Behaviour, Fate and Transport of Two Diluted Bitumen Products from the Canadian Oil Sands*; Cat. No.: En84-96/2013E-PDF; Government of Canada: Ottawa, ON, Canada, 2013; pp. 1–85.
11. King, T.L.; Robinson, B.; Boufadel, M.; Lee, K. Flume tank studies to elucidate the fate and behavior of diluted bitumen spilled at sea. *Mar. Pollut. Bull.* **2014**, *83*, 32–37. [CrossRef] [PubMed]
12. Cole, M.G.; King, T.L.; Lee, K. *Analytical Technique for Extracting Hydrocarbons from Water Using Sample Container as Extraction Vessel in Combination with Roller Apparatus*; Canadian Technical Report of Fisheries and Aquatic Sciences #2733; Fisheries and Oceans Canada: Québec, QC, Canada, 2007; pp. 1–12.
13. Stevens, C.C.; Thibodeaux, L.J.; Overton, E.B.; Valsaraj, K.T.; Nandakumar, K.; Rao, A.; Walker, N.D. Sea surface oil slick light component vaporization and heavy residue sinking: binary mixture theory and experimental proof of concept. *Environ. Eng. Sci.* **2015**, *32*, 694–702. [CrossRef]
14. King, T.; Robinson, B.; McIntyre, C.; Toole, P.; Ryan, S.; Saleh, F.; Boufadel, M.; Lee, K. Fate of surface spills of Cold Lake blend diluted bitumen treated with dispersant and mineral fines in a wave tank. *Environ. Eng. Sci.* **2015**, *32*, 250–261. [CrossRef]
15. Baldwin, I.T.; Staszak-Kozinski, L.; Davidson, R. Up in smoke: I. Smoke-derived germination cues for postfire annual, Nicotiana attenuata Torr. Ex. Watson. *J. Chem. Ecol.* **1994**, *20*, 2345–2371. [PubMed]
16. Arkles, B. Hydrophobicity, hydrophilicity and silanes. *Paint Coat. Ind. Mag.* **2006**, *22*, 114–123.
17. Coates, J. *Interpretation of Infrared Spectra: A Practical Approach. Encyclopedia of Analytical Chemistry*; Meyers, R.A., Ed.; John Wiley & Sons Ltd.: Chichester, UK, 2000; pp. 10815–10837.

Journal of
Marine Science and Engineering

MDPI

Article

Human Genotoxic Study Carried Out Two Years after Oil Exposure during the Clean-Up Activities Using Two Different Biomarkers

Gloria Biern [1], Jesús Giraldo [2], Jan-Paul Zock [3,4,5], Gemma Monyarch [1], Ana Espinosa [3,4,5], Gema Rodríguez-Trigo [6,7,8], Federico Gómez [8,9], Francisco Pozo-Rodríguez [8,10], Joan-Albert Barberà [8,9] and Carme Fuster [1,*]

[1] Unitat de Biologia Cel·lular i Genètica Mèdica, Facultat de Medicina, Universitat Autònoma de Barcelona (UAB), 08193-Bellaterra, Barcelona, Spain; gloria.biern@gmail.com (G.B.); taguca@gmail.com (G.M.)

[2] Unitat de Bioestadística i Institut de Neurociències, Facultat de Medicina, Universitat Autònoma de Barcelona (UAB), 08193-Bellaterra, Barcelona, Spain; jesus.giraldo@uab.es

[3] Centre de Recerca en Epidemiologia Ambiental (CREAL), 08003-Barcelona, Spain; jpzock@creal.cat (J.-P.Z.); aespinosa@creal.cat (A.E.)

[4] Universitat Pompeu Fabra, 08002-Barcelona, Spain

[5] CIBER Epidemiología y Salud Pública (CIBERESP), 28029-Madrid, Spain

[6] Servicio de Neumología, Hospital Clínico San Carlos, 28040-Madrid, Spain; grodriguezt@salud.madrid.org

[7] Facultad de Medicina, Universidad Complutense, 28040-Madrid, Spain

[8] CIBER Enfermedades Respiratorias (CIBERES), Bunyola, 07004-Mallorca, Spain; fpgomez@clinic.ub.es (F.G.); fpozo@h12o.es (F.P.-R.); jbarbera@clinic.ub.es (J.-A.B.)

[9] Departamento de Medicina Respiratòria, Hospital Clínic-Institut d'Investigacions Biomèdiques August Pi I Sunyer (IDIBAPS), 08036-Barcelona, Spain

[10] Departamento de Medicina Respiratoria, Unidad Epidemiologia Clínica, Hospital 12 de Octubre, 28047-Madrid, Spain

[*] Author to whom correspondence should be addressed; carme.fuster@uab.es; Tel.: +34-93-581-1273; Fax +34-93-581-1025.

Academic Editor: Merv Fingas
Received: 22 September 2015; Accepted: 27 October 2015; Published: 3 November 2015

Abstract: Micronuclei, comet and chromosome alterations assays are the most widely used biomarkers for determining the genotoxic damage in a population exposed to genotoxic chemicals. While chromosome alterations are an excellent biomarker to detect short- and long-term genotoxic effects, the comet assay only measures early biological effects, and furthermore it is unknown whether nuclear abnormalies, such as those measured in the micronucleus test, remain detectable long-term after an acute exposure. In our previous study, an increase in structural chromosome alterations in fishermen involved in the clean-up of the *Prestige* oil spill, two years after acute exposure, was detected. The aim of this study is to investigate whether, in lymphocytes from peripheral blood, the nuclear abnormalies (micronucleus, nucleoplasmic bridges and nuclear buds) have a similar sensitivity to the chromosome damage analysis for genotoxic detection two years after oil exposure in the same non-smoker individuals and in the same peripheral blood extraction. No significant differences in nuclear abnormalies frequencies between exposed and non-exposed individuals were found ($p > 0.05$). However, chromosome damage, in the same individuals, was higher in exposed *vs.* non-exposed individuals, especially for chromosome lesions ($p < 0.05$). These findings, despite the small sample size, suggest that nuclear abnormalities are probably less-successful biomarkers than are chromosome alterations to evaluate genotoxic effects two or more years after an exposure to oil. Due to the great advantage of micronucleus automatic determination, which allows for a rapid study of hundreds of individuals exposed to genotoxic chemical exposure, further studies are needed to confirm whether this assay is or is not useful in long-term genotoxic studies after the toxic agent is no longer present.

J. Mar. Sci. Eng. **2015**, *3*, 1334–1348

Keywords: micronucleus test; chromosome damage; nuclear abnormities; chromosome alterations; oil exposure; genotoxicity; *Prestige* catastrophe

---

1. Introduction

Significant marine oil spills, approximately namely 14 accidents involving large oil tankers, have occurred in regions with a high population density in the last five decades [1,2]. When a sizeable spill occurs, there are usually a large number of individuals, in general local inhabitants, who collaborate in clean-up tasks to minimize the negative ecological and economic impact. So, for example, more than 300,000 people were involved in the clean-up activities after the wreck of the oil tanker *Prestige*, in November 2002. Although there are relatively few studies which have focused on the repercussions of acute oil exposure for human health, there is growing concern about the chemical exposure that clean-up activities involve and their potential health effects. Direct contact with oil or its vapors can cause skin rash and eye redness, and prolonged exposure can cause nausea, dizziness, headache, respiratory problems and psychiatric disorders [1,2]. Moreover, due to certain volatile organic oil compounds, in particular benzene, being carcinogenic in humans [3], it is very important to determine if exposure to oil during clean-up tasks is associated with genotoxic effects in the short- (less than 12 months) and long-term (more than one year). So far, only a few human genotoxic studies in oil exposed populations have been published, most after the wreck of the *Prestige* [4–14]. In general, these studies revealed increased genomic damage in exposed individuals during the clean-up tasks [4–11]. Nevertheless, only two research groups have carried out long-term genotoxic studies after oil exposure to the *Prestige* [12–15], with conflicting results. In one group, the authors described an increase of structural chromosome alterations in highly exposed *vs.* non-exposed individuals two and six years after exposure [12–14], this follow-up study reveal that chromosome damage persisted at least for the six years. Yet in another group, the study detected no genotoxic effects to be present seven years after exposure using other biomarkers (comet, micronucleus and T-cell receptor mutation assays) [15]. With the exception of T-cell receptor mutation assays, the sensibility of the two other biomarkers to detect long-term genotoxic effects has not been tested.

The micronucleus test, comet assay and chromosome alterations have been the most common biomarkers to determine genetic damage in any population exposed occupationally or environmentally to genotoxic chemicals, e.g., oil exposure during clean-up tasks [4–18]. A micronucleus is the result of chromosome breakage (acentric fragment) and/or loss (whole chromosome) caused by errors in DNA repair or in chromosome segregations not included in the main nucleus that are surrounded individually by the nuclear membrane [19]. The micronucleus test is performed by cytokinesis-block assays using cytochalasin B, which allows to be analyzed other nuclear abnormalies, such as nucleoplasmic bridges and nuclear buds in binuclear cells. The nucleoplasmic bridges indicate the occurrence of reorganizations in which chromatids or chromosomes are pulled to opposite poles during anaphase, resulting in dicentric or ring chromosomes. The nuclear buds are characterized by having the same morphology as a micronucleus, but they remain connected to the main nucleus and represent the process of elimination of amplified DNA or of the DNA repair complex and possibly excess chromosomes from aneuploid cells. Recently, it has been described that nucleoplasmic bridges and nuclear buds are also useful biomarkers for monitoring genetic damage by detecting and quantifying DNA damage and chromosome instability [20–23]. The comet biomarker is based on how a genotoxic agent will produce DNA-strand breaks and measures the extent of DNA migration in electrophoresis [24] and has been frequently used because it is a fast and easy method to assess DNA breaks with excellent sensitivity. Finally, chromosome alterations are any change in the normal structure or number of chromosomes. In general, their analyses for genotoxic studies include lesions (gaps and breaks of one or both chromatids) and structural alterations (such as acentric fragments, deletions, translocations, dicentrics, rings, marker chromosomes) resulting in direct DNA breakage, errors in synthesis or repair of DNA, and have been widely used biomarkers since the 1970s [16,23,24]. Although comet and nuclear anomaly assays are probably less resolute and less informative than is metaphasic chromosome analysis, in the last several years both tests have been used extensively in studies to evaluate genotoxic damage in large populations of exposed individuals because they are much easier and faster [19].

The evaluation of chromosome alterations requires cell cultures, while the evaluation of nuclear abnormalities requires cells in division but comet assay can be performed without the use of proliferative cells. For a long time, peripheral lymphocyte cultures have been the most widely employed in human genotoxic studies, however the introduction of nuclear abnormalities and comet assays as biomarkers allows for the use of alternative cell types, such as epithelial cells [25,26]. Epithelial cells can be used as early-effect biomarkers; nasal epithelial cells are replaced approximately once every 30 days and buccal epithelial cells one every 10–14 days [26] *vs.* peripheral blood lymphocytes, which serve as long-effect biomarkers and are renewed around every 4 to 6 years [27]. The great advantage of using, for example, buccal epithelial cells *vs.* lymphocytes is the easy and minimally invasive collection of samples, but the most important disadvantages are the discrepancies which come from using blood cells.

In contrast to the body of research regarding the use of different biomarkers to determine the genotoxic effect when the agent is present, there is scarce information to determine long-term effects after an acute exposure, with chromosome damage being the biomarker most frequently used. Given that the comet test indicates early biological effects [28,29]; it is probably not an ideal biomarker for long-term studies after acute genotoxic exposure. Although the usefulness of nuclear abnormalies, especially the micronucleus test, for short-term genotoxic studies is unquestionable, its sensibility for long-term studies, when the toxic agent is missing, has not been demonstrated yet. The main objective of this study is to determine whether nuclear abnormalies remain useful biomarkers for detecting genotoxic effects two years after Prestige oil exposure, comparing their results with those detected by chromosome alterations analyses.

2. Material and Methods

2.1. Study Population

The present study was performed on randomly selected subsamples of individuals included in a previous genotoxic study [12,13]. It was conducted using peripheral blood lymphocytes from individuals who had participated in clean-up activities of the *Prestige* oil spill. Only fishermen were included in our study in order to minimize other occupational sources of exposure that could act as confounders. A questionnaire survey including information about participation in clean-up tasks, health problems, lifestyle, history of cancer, medication, smoking status, fertility, age, and gender among 6780 fishermen one year after exposure was performed [30]. The selection criteria of individuals highly exposed and non-exposed to the oil was established from this information, described previously [30]. In brief, exposed local fishermen who participated for at least 15 days in clean-up activities of the *Prestige* oil spill, for four or more hours per day, during the first two months (when exposure presumably was greatest) were included as highly-exposed subjects for the study. Non-exposed individuals were selected from fishermen who had not participated in clean-up tasks for reasons other than those related to health. All exposed and non-exposed individuals were non-smokers (current smokers and ex-smokers were excluded), fertile and without a history of cancer, A new questionnaire and face-to-face interview, in order to verify the answers, was performed in a mobile unit that traveled to the participants' villages on the same day in which the samples were obtained two years after the spill. In the present work, a total of 20 exposed and 20 non-exposed individuals were studied, randomly selected from 91 exposed and 46 non-exposed individuals in which chromosome damage was analyzed [12,13]. Figure 1 shows the flow diagram of the study. The exposed group consisted of 9 men and 11 women with an average age of 48.2 years (ranging from 32.2 to 62.2; SEM = 1.9). The non-exposed group consisted of 3 men and 17 women with an average

age of 53.1 years (ranging from 36.6 to 58.8; SEM = 1.3). No significant relationship between sex and group was found according to Fisher's exact test ($p = 0.0824$). The difference in age was found to be statistically significant both by Student's t-test and Wilcoxon sum of ranks. In principle, as a higher age could be associated with a higher propensity to present genotoxic abnormalities, the distribution of age in the sample could make it more difficult to statistically prove the association between abnormalities and oil exposure. Thus, because there are more older individuals included in the non-exposed group, age distribution should not favor the hypothesis of finding abnormalities in the exposed group. The collection, transport and processing of the samples were performed between 22 and 27 months after the *Prestige* disaster.

The project was approved by the Ethics Committee on Clinical Research of Galicia, and all participants provided written, informed consent.

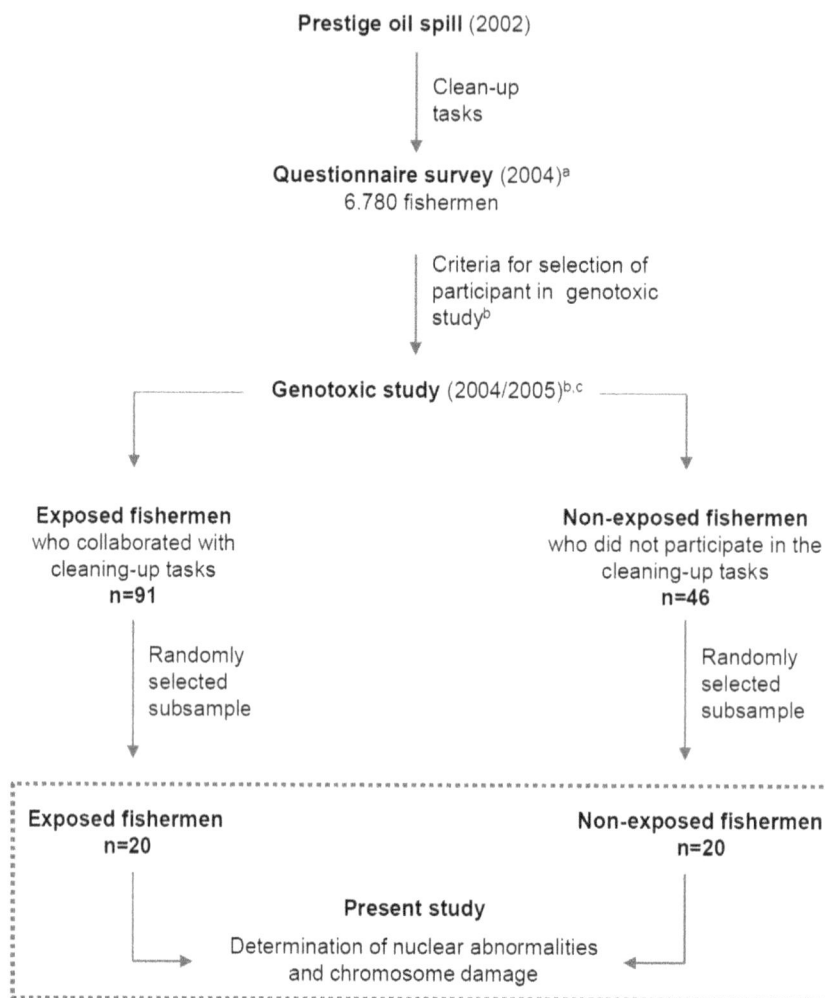

Prestige oil spill (2002)

| Clean-up tasks

Questionnaire survey (2004)[a]
6.780 fishermen

| Criteria for selection of participant in genotoxic study[b]

Genotoxic study (2004/2005)[b,c]

Exposed fishermen who collaborated with cleaning-up tasks n=91	**Non-exposed fishermen** who did not participate in the cleaning-up tasks n=46
Randomly selected subsample	Randomly selected subsample
Exposed fishermen n=20	**Non-exposed fishermen** n=20

Present study
Determination of nuclear abnormalities and chromosome damage

Figure 1. Flow diagram of the study. [a] Detailed description in Zock *et al.* [30]; [b] Detailed description in Rodriguez-Trigo *et al.* [12]; and [c] Detailed description in Monyarch *et al.* [13].

2.2. Cytogenetic Analysis

Peripheral blood was obtained in same extraction and later cultured at 37 °C in supplemented RPMI-1640 medium (GIBCO Invitrogen Cell Culture, Invitrogen; Carlsbad, CA, USA) according to standard procedures.

For the cytokinesis-block nuclear abnormalies test, peripheral blood was cultured, in duplicate, for 44 h and then cytochalasin-B was added to a final concentration of 6 µg/mL. Cells were harvested by centrifugation after 72 h of culture and submitted to middle-hypotonic treatment with 0.075 mM KCl at 4 °C. Cells were fixed in Carnoy (methanol-acetic acid 3:1 v/v), placed on dry slides, and stained with Leishman according to standard procedures. The micronucleus, nucleoplasmic bridges and nuclear buds in binucleated cells were identified according to the criteria of the HUMN project [31] and were evaluated by scoring 1000 binucleated cells (500 from each culture) using an Olympus Bx60 microscope. The cytogenetic-block proliferation index was calculated by the relation between total of cells with 1, 2, 3 and 4 micronuclei *vs.* total of cells analyzed.

For chromosome breakage analyses involved in lesions and structural chromosome alterations, analyzed in published studies [12,13], peripheral blood was cultured, in duplicate, for 72 h and then harvested according to standard procedures. For chromosome lesions, a minimum at 100 metaphases were analyzed in each individual (50 from each culture). For structural chromosome alterations, at least 25 banded metaphases were karyotyped in each participant. Criteria for cytogenetic evaluations were determined according to the International System for Human Cytogenetic Nomenclature [32].

2.3. Statistical Analysis

A generalized estimating equation, GEE [12,13,33,34], was used for assessing the differences between the exposed and non-exposed groups for the micronucleus, nucleoplasmic bridges and nuclear buds, chromosome lesions and structural chromosome alterations. The GEE approach is an extension of generalized linear models designed to account for repeated, within-individual measurements. This method is particularly indicated for when the normality assumption is not reasonable, as happens, for instance, with discrete data. The GEE model was used instead of the classic Fisher exact test because the former takes into account the possible within-individual correlation, whereas the latter assumes that all observations are independent. Since several metaphases were analyzed per individual, the GEE model is more appropriate. Statistical significance was set at $p < 0.05$. Statistical analyses were carried out with SAS/STAT release 9.02 (SAS Institute Inc.; Cary, NC, USA). The GEE model was fitted using the REPEATED statement in the GENMOD procedure. The conservative Type 3 statistics score was used for the analysis of the effects in the model. Sex and age were found not to be statistically significant when included in the GEE model and therefore were removed from the analysis.

3. Results

A satisfactory cell growth in all cultures was observed. A total of 40,000 binucleate cells, 4260 metaphases and 1100 karyotypes were analyzed in lymphocytes from exposed and non-exposed individuals respectively. All individuals had normal karyotypes (46,XX or 46,XY), except two cases, one with a polymorphic inversion of chromosome 9, inv(9)(p11q12), in an exposed individual (E14) and another case with an increased length of the heterochromatin on the long arm of the Y chromosome, Yqh+, in a non-exposed individual (NE5).

Cell growth in cytochalasin-B cultures (from exposed and non-exposed individuals) showed a cytogenetic-block proliferation index ranging between 30% and 60%. No significant statistical differences were found in the micronucleus or nuclear buds between exposed and non-exposed individuals ($p = 0.4774$ and $p = 0.2356$, respectively), and nucleoplasmic bridges were marginally influential ($p = 0.08$).

Chromosome lesions were higher in exposed rather than in non-exposed individuals ($p = 0.0231$), but structural chromosome alterations were only marginally ($p = 0.0972$). Marker chromosomes, unbalanced translocations and deletions were the structural chromosome alterations most frequently observed in both groups of individuals, and ring chromosomes and acentric fragments were only detected in exposed individuals. Numerical chromosome alterations (such as trisomies and monosomies) were excluded in these analyses because they can be attributed to the failure of the chromosome spread due to non-specific techniques having been applied to remaining cell membranes.

Table 1 and Figure 2 show the nuclear abnormalies (micronucleus, nucleoplasmic bridges and nuclear buds) and chromosome damage (lesions and structural alterations) observed in the same individuals. Cytogenetic results for each individual are found in Table 2, showing high inter-individual genotoxic variability for all biomarkers analyzed (micronucleus, nucleoplasmic bridges, nuclear buds, chromosome lesions and structural chromosome alterations) in exposed and in non-exposed individuals. With the exception of the degree of oil exposure no other associations were found between genotoxic damage and the different factors analyzed in the present study.

Table 1. Nuclear abnormalities and chromosome damage detected in same individuals exposed and non-exposed to oil.

	Exposed	Non-Exposed	*p*-Value
Total Individuals, No.	20	20	
Total Binucleate Cells, No.	20.000	20.000	
Binucleated cells with micronucleus, No. (%)	457 (2.3)	514 (2.6)	0.4774
1 micronucleus/cell, No.	399	450	
2 micronucleus/cell, No.	49	53	
3 micronucleus/cell, No.	9	11	
Nucleplasmic bridges, No. (%)	131 (0.65)	98 (0.49)	0.08
Nuclear buds, No. (%)	106 (0.53)	68 (0.34)	0.2356
Total Metaphases Analyzed (Uniform Stain), No.	2112	2.148	
Chromosome lesion, No. (%)	28 (1.3)	7 (0.3)	0.0231
Total Metaphases Karyotyped (G-Banded), No.	537	563	
Structural chromosome alterations, No. (%)	36 (6.7)	16 (2.8)	0.0972
Balanced, No.	1	3	
Unbalanced, No.	35	13	

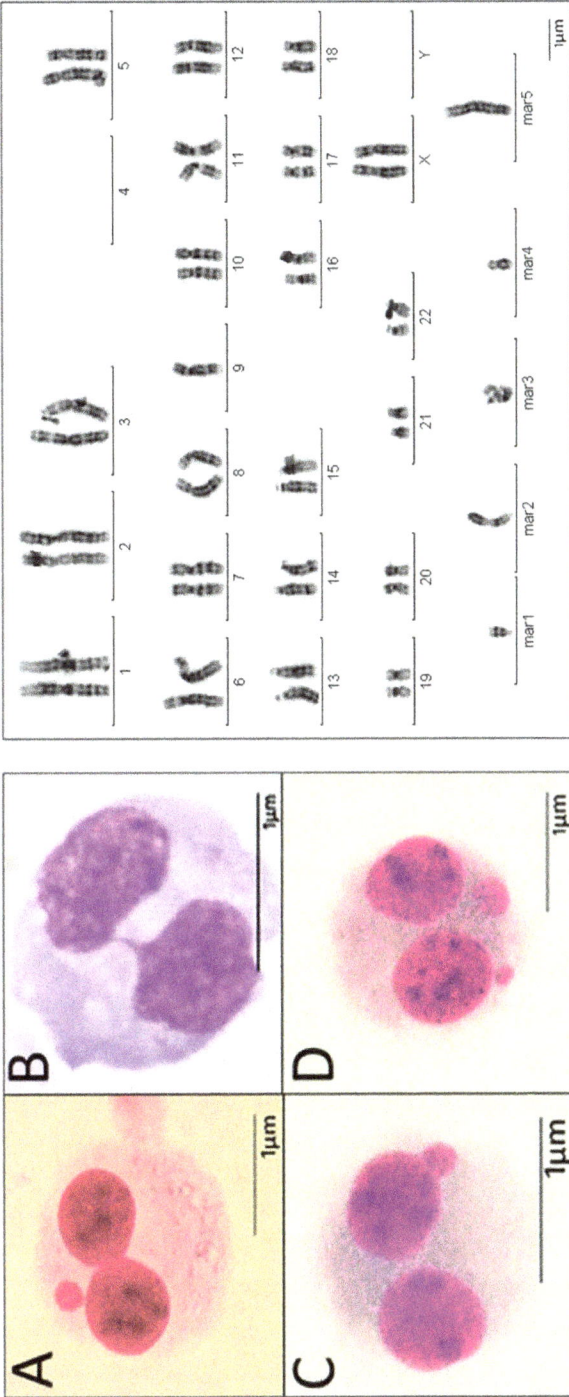

Figure 2. Genotoxic study using nuclear abnormalities and chromosome damage as biomarkers. Binuclear cells (left) showing: (**A**) one micronucleus, (**B**) nucleoplasmatic bridges, (**C**) nuclear buds, (**D**) two micronuclei; and Karyotype (right) with five marker chromosomes.

Table 2. Cytogenetic results in individuals exposed and non-exposed to oil.

Type of individuals	Sex	Age	Cells (MN)	MN 0	MN 1	MN 2	MN ≥3	NBUP	NPB	Cells	Lesions	Karyotypes	Structural Alterations	Balanced Structural Alterations	Unbalanced Structural Alterations	Type of Structural Alteration
Exposed																
E1	woman	49.02	1000	29	971	27	2	0	4	7	100	1	27	5	0	5	t(13;20)(q32;q12); t(9;7), t(10;18),mar; del(2)(q12)
E2	man	54.01	1000	23	977	20	3	0	2	4	116	3	28	0	0	0	
E3	man	44.02	1000	21	979	19	2	0	5	6	105	6	29	1	0	1	ace
E4	man	31.26	1000	17	983	16	1	0	5	1	105	2	25	2	0	2	ace,ace
E5	woman	51.90	1000	25	975	22	3	0	11	8	103	1	26	0	0	0	
E6	man	35.92	1000	17	983	16	1	0	3	8	108	1	30	2	0	2	ace; add(12)(qter)
E7	woman	49.62	1000	43	957	37	4	2	9	5	100	2	25	6	0	6	ace; t(4;16)(q13,p13.3); mar,mar,mar,mar
E8	man	56.45	1000	43	957	37	4	2	19	6	102	1	29	0	0	0	
E9	woman	50.51	1000	11	989	9	2	0	16	14	100	6	26	1	0	1	ace
E10	man	52.50	1000	9	991	8	1	0	2	17	102	1	26	0	0	0	
E11	woman	54.86	1000	32	968	24	8	0	4	5	110	0	25	1	0	1	mar
E12	woman	44.89	1000	37	963	30	5	2	12	2	112	1	26	1	1	0	t(13;11)(p25;q23)
E13	man	57.02	1000	18	982	18	0	0	9	3	100	1	31	1	0	1	mar
E14	man	46.13	1000	18	982	16	1	1	7	3	104	1	26	1	0	1	t(7;10); r, r
E15	woman	62.17	1000	14	986	11	3	0	4	3	103	0	25	3	0	3	t(X;4)(q21;p16)
E16	woman	48.99	1000	10	990	8	2	0	1	1	112	0	25	1	0	1	del(7)(q33)
E17	woman	58.37	1000	8	992	8	0	0	1	3	105	0	27	1	0	1	mar
E18	woman	36.90	1000	31	969	26	3	2	4	4	112	0	30	1	0	1	del(7)(p15)
E19	man	52.98	1000	38	962	34	4	0	5	6	106	2	26	1	0	1	del(7)(p15)
E20	man	54.94	1000	13	987	13	0	0	6	3	107	1	25	9	1	9	mar,mar; ace, ace; mar, mar,mar,mar,mar
			20000	457	19543	399	49	9	131	106	2112	28	537	36	1	35	36
Non-Exposed																
NE1	woman	54.33	1000	22	978	19	3	0	3	0	103	0	25	0	0	0	
NE2	woman	50.60	1000	22	978	17	4	1	0	0	100	3	34	0	0	0	
NE3	woman	57.52	1000	13	987	10	2	1	4	2	105	1	25	4	0	4	mar,mar,mar,mar
NE4	woman	57.19	1000	14	986	13	1	0	16	0	101	1	25	1	0	1	del(1)(q21)
NE5	man	58.78	1000	22	978	19	2	1	1	8	108	0	26	3	1	3	del(2)(q21); del(1)(q23), mar
NE6	woman	57.61	1000	59	941	52	6	1	2	6	100	0	25	1	1	1	t(13;14)(q14,q32)
NE7	woman	36.56	1000	5	995	3	2	0	5	7	107	0	25	1	0	1	del(1)(q32)
NE8	woman	53.22	1000	30	970	26	3	1	7	7	110	0	33	0	0	0	mar
NE9	woman	58.58	1000	50	950	43	6	1	5	1	113	0	43	0	0	0	
NE10	woman	56.00	1000	26	974	22	3	1	9	5	107	1	26	0	0	0	
NE11	woman	56.85	1000	10	990	10	0	0	3	4	120	0	25	1	0	1	mar
NE12	woman	46.36	1000	38	962	32	5	1	9	5	130	0	26	0	0	0	
NE13	woman	55.82	1000	12	988	10	2	0	2	6	101	0	25	0	0	0	
NE14	woman	45.84	1000	26	974	24	2	0	8	4	108	2	25	0	0	0	
NE15	man	48.56	1000	41	959	33	7	1	8	3	107	0	25	0	0	0	
NE16	woman	55.54	1000	23	977	23	0	0	5	2	102	0	26	0	0	0	
NE17	woman	58.62	1000	36	964	32	4	0	8	4	112	0	26	1	0	1	del(9)(q21)
NE18	man	57.15	1000	38	962	37	1	0	6	3	105	0	43	0	0	0	
NE19	woman	49.98	1000	18	982	16	2	0	3	1	108	0	26	2	0	1	t(8;13)(q24.1;q31); t(2;5)
NE20	woman	46.31	1000	9	991	9	0	0	0	4	101	0	29	1	1	1	t(3;9)(q27;q13)
			20000	514	19486	450	53	11	98	68	2148	7	563	16	3	13	16

Abbreviations: ace: acentric fragment; add: additional material of unknown origin; del: deletion; mar: marker chromosome; p: short arm; q: long arm; qter: terminal long arm; t: translocation; MN: micronuclei; NBUP: nucleoplasmic bridges; NPB: nuclear buds. Commas indicate the beginning of new metaphase.

4. Discussion

Toxic agents can induce complex changes in the genome, and to-date there is no single biomarker to detect all types of these alterations, probably due to different molecular mechanisms being involved [29,35,36]. It is therefore probable that not all genotoxic biomarkers are equally useful for long-term evaluation after exposure.

To date, very few long-term genotoxic studies after an accidental oil exposure have been previously published [12–15]. In all of these studies only individuals highly exposed to oil were included, yet the findings obtained were not coincident. While an increase of structural chromosome alterations in exposed individuals two and six years after exposure was observed [12–14], no genotoxic effects using other biomarkers (comet, micronucleus and T-cell receptor mutation assays) after seven years were detected [15]. The T-cell receptor mutation assay, used in Laffon's study [15], is an excellent biomarker for long-term studies because it provides information about the genotoxic effect which have occurred several months to several years after exposure [37]. However, the comet assay, also a successful biomarker employed by the authors, indicates early biological effects [28,29], so is probably not the most suitable test to evaluate long-term genotoxic effects. In relation to nuclear abnormalies, with the exception of Laffon *et al.* [15] and the present study, no other long-term genotoxic analyses have been performed, and in both studies no genotoxicity was detected two and seven years after oil exposure. Thus, we are not sure that this biomarker is still valid for long-term analysis.

The main differences in the above referred studies [12–15] were the individuals included in the study, the time following oil exposure (two, six and seven years), and the biomarkers used (mainly chromosome damage *vs.* micronuclei and nuclear abnormalies). In order to minimize the dispersion of these factors, we have analyzed these same biomarkers two years after oil exposure in the same individuals in which chromosome alterations were observed. Our results showed no differences in micronuclei and nuclear abnormalies between those exposed and non-exposed to oil. Thus, if we had only used these biomarkers, our findings would have suggested that the genotoxic effect has disappeared two years after oil exposure, long before the seven years as described Laffon *et al.* [15] However, the present study shows an increase of chromosome damage in the same exposed individuals, in which no differences for nuclear abnormalies were found, indicating that genotoxic damage does persist two years after acute oil exposure. Additionally, an increase of chromosome damage in a high number of exposed individuals was previously reported two and six years after oil exposure [12–14]. Despite the very strict selection criteria for exposed individuals and the small sample size analyzed, the present findings, supported by those reported previously [12–15], suggest that micronuclei and nuclear abnormalies are probably less-successful biomarkers than are chromosome damage to evaluate genotoxic effect more than two years after acute oil exposure when the toxic agent is no longer present. Moreover, our results indicate that chromosome damage is more informative than micronuclei and nuclear abnormalies because acentric fragments (which corresponds to the origin of the micronucleus) and ring chromosomes (corresponding to nucleoplasmic bridges) were detected in exposed individuals two years after oil exposure. It is relevant to note that smokers were excluded in present study because an association between smoking and chromosome damage has been described [33,38]. Finally, due to the fact that the micronucleus test and other nuclear abnormalies can be detected automatically *versus* chromosome analysis, and moreover this test is much easier, faster and allows for analysis of a large number of cells and does not require as much extensive personnel training, further studies are needed to confirm these preliminary results in larger samples.

5. Conclusions

To date, no information has been published regarding whether the micronucleus test remains suitable several years after the toxic agent is no longer present. For this reason, in the present study we evaluated the utility of nuclear abnormalies, including micronucleus test, to assess the genotoxic oil effect two years after the wreck of the *Prestige* comparing these results with those obtained from chromosome alterations analyses in the same non-smoker individuals and in the same peripheral blood extraction. Our results showed no differences in nuclear abnormalies between those exposed and non-exposed, however the chromosome damage was higher in exposed individuals. These features were compared with previous a report derived from long-term genotoxic studies after an accidental oil exposure. The main findings are:

- nuclear abnormalies (micronucleus, nucleoplasmic bridges and nuclear buds) in binucleated cells may not detect genotoxic effects more than two years after acute oil exposure when the toxic agent is no longer present;
- chromosome damage (chromosome lesion and structural chromosome alterations) in metaphases cells is a useful biomarker for assessing genotoxic effect two years after acute oil exposure using the same peripheral blood extraction in which nuclear abnormalies were analyzed; and
- comparative study using nuclear abnormalies and chromosome damage analyses emphasizes the need to use appropriate biomarker for detection of genotoxic effect in individuals involved in toxic accidents.

Despite the reduced number of individual analyzed, the present study suggests that micronuclei and nuclear abnormalies are probably less-successful biomarkers for the evaluation of long-term genotoxic oil effects when the toxic agent is no longer present. However, due to the fact that with the micronucleus test these and other nuclear abnormalies can be detected automatically, further studies are needed to confirm these preliminary results.

Acknowledgments: The kind participation of the fishermen's cooperatives and the efforts made by Antonio Devesa are gratefully acknowledged. The authors wish to thank Ana Souto Alonso, Marisa Rodríguez Valcárcel, Luisa Vázquez Rey, Emma Rodríguez (Complexo Hospitalario Universitario A Coruña), and Ana Utrabo, Angels Niubó (Universitat Autònoma de Barcelona). The investigators are greatly indebted to J. Ancochea and J.L. Alvarez-Sala, past presidents of SEPAR, for their initiative and support. For this study, grants were provided from the Health Institute Carlos III FEDER (PI03/1685), Sociedad Española de Neumología y Cirugía Torácica (SEPAR), Cowmissionat per a Universitats I Recerca from Generalitat de Catalunya (14SGR903), Centro de Investigación en Red de Enfermedades Respiratorias (CIBERES). The different sponsors were not involved in the design of the study, sample collection, cytogenetic analysis, or interpretation of the data or preparation/revision of the manuscript.

Author Contributions: Conception and design of the experiments: C.F. Analysis and interpretation of the data: G.B., G.M., C.F. Statistical analysis: J.G., A.E., J.P.Z. Composition and revision of the manuscript: J.G., J.P.Z., G.R.T., F.P.G., F.P.R., J.A.B. and C.F.

Conflicts of Interest: The authors declare no conflict of interest.

References

1. Aguilera, F.; Méndez, J.; Pásaro, E.; Laffon, B. Review of the effects of exposure to spilled oils on human health. *J. Appl. Toxicol.* **2010**, *30*, 291–301. [CrossRef] [PubMed]
2. Goldstein, B.D.; Osofsky, H.J.; Lichtveld, M.Y. The Gulf oil spill. *N. Engl. J. Med.* **2011**, *364*, 1334–1348. [CrossRef] [PubMed]
3. IARC. *Occupational Exposures in Petroleum Refining: Crude Oil and Major Petroleum Fuels IARC. Monographs on the Evaluations of Carcinogenic Risk to Humans;* International Agency for Research on Cancer: Lyon, France, 1989; Volume 45.
4. Clare, M.G.; Yardley-Jones, A.; Maclean, A.C.; Dean, B.J. Chromosome analysis from peripheral blood lymphocytes of workers after an acute exposure to benzene. *Br. J. Ind. Med.* **1984**, *41*, 249–253. [CrossRef] [PubMed]
5. Cole, J.; Beare, D.M.; Waugh, A.P.; Capulas, E.; Aldridge, K.E.; Arlett, C.F.; Green, M.H.; Crum, J.E.; Cox, D.; Garner, R.C.; *et al.* Biomonitoring of possible human exposure to environmental genotoxic chemicals: Lessons from a study following the wreck of the oil tanker Braer. *Environ. Mol. Mutagen.* **1997**, *30*, 97–111. [CrossRef]
6. Laffon, B.; Fraga-Iriso, R.; Perez-Cadahia, B.; Méndez, J. Genotoxicity associated to exposure to Prestige oil during autopsies and cleaning of oil-contaminated birds. *Food Chem. Toxicol.* **2006**, *44*, 1714–1723. [CrossRef] [PubMed]

7. Perez-Cadahia, B.; Laffon, B.; Pasaro, E.; Méndez, J. Genetic damage induced by accidental environmental pollutants. *Sci. World J.* **2006**, *6*, 1221–1237. [CrossRef] [PubMed]

8. Pérez-Cadahía, B.; Lafuente, A.; Cabaleiro, T.; Pasaro, E.; Méndez, J.; Laffon, B. Initial study on the effects of Prestige oil on human health. *Environ. Int.* **2007**, *33*, 176–185. [CrossRef] [PubMed]

9. Pérez-Cadahía, B.; Laffon, B.; Porta, M. Relationship between blood concentrations of heavy metals and cytogenetic and endocrine parameters among subjects involved in cleaning coastal areas affected by the Prestige tanker oil spill. *Chemosphere* **2008**, *7*, 447–455. [CrossRef] [PubMed]

10. Pérez-Cadahía, B.; Laffon, B.; Valdiglesias, V.; Pasaro, E.; Mendez, J. Cytogenetic effects induced by Prestige oil on human populations: The role of polymorphisms in genes involved in metabolism and DNA repair. *Mutat. Res.* **2008**, *653*, 117–123. [CrossRef] [PubMed]

11. Perez-Cadahia, B.; Mendez, J.; Pasaro, E.; Lafuente, A.; Cabaleiro, T.; Laffon, B. Biomonitoring of human exposure to Prestige oil: Effects on DNA and endocrine parameters. *Environ. Health Insights* **2008**, *2*, 83–92. [PubMed]

12. Rodríguez-Trigo, G.; Zock, J.P.; Pozo-Rodríguez, F.; Gómez, F.P.; Monyarch, G.; Bouso, L.; Coll, M.D.; Verea, H.; Antó, J.M.; Fuster, C.; *et al.* SEPAR-Prestige Study Group. Health changes in fishermen 2 years after clean-up of the Prestige oil spill. *Ann. Intern. Med.* **2010**, *153*, 489–498. [CrossRef] [PubMed]

13. Monyarch, G.; de Castro-Reis, F.; Zock, J.P.; Giraldo, J.; Pozo-Rodríguez, F.; Espinosa, A.; Rodríguez-Trigo, G.; Verea, H.; Castaño-Vinyals, G.; Gómez, F.P.; *et al.* Chromosomal bands affected by acute oil exposure and DNA repair errors. *PLoS ONE* **2013**, *8*, e81276. [CrossRef] [PubMed]

14. Hildur, K.; Templado, C.; Zock, J.P.; Giraldo, J.; Pozo-Rodríguez, F.; Frances, A.; Monyarch, G.; Rodríguez-Trigo, G.; Rodriguez-Rodriguez, E.; Souto, A.; *et al.* Follow-up genotoxic study: Chromosome damage two and six years after exposure to the *Prestige* oil spill. *PLoS ONE* **2015**, *10*. [CrossRef]

15. Laffon, B.; Aguilera, F.; Ríos-Vázquez, J.; Valdiglesias, V.; Pásaro, E. Follow-up study of genotoxic effects in individuals exposed to oil from the tanker Prestige, seven years after the accident. *Mutat. Res.* **2014**, *760*, 10–16. [CrossRef] [PubMed]

16. Mateuca, R.; Lombaert, N.; Aka, P.V.; Decordier, I.; Kirsch-Volders, M. Chromosomal changes: Induction, detection methods and applicability in human biomonitoring. *Biochimie* **2006**, *88*, 1515–1531. [CrossRef] [PubMed]

17. Valverde, M.; Rojas, E. Environmental and occupational biomonitoring using the Comet assay. *Mutat. Res.* **2009**, *681*, 93–109. [CrossRef] [PubMed]

18. DeMarini, D.M. Genotoxicity biomarkers associated with exposure to traffic and near-road atmospheres: A review. *Mutagenesis* **2013**, *28*, 485–505. [CrossRef] [PubMed]

19. Fenech, M.; Kirsch-Volders, M.; Rossnerova, A.; Sram, R.; Romm, H.; Bolognesi, C.; Ramakumar, A.; Soussaline, F.; Schunck, C.; Elhajouji, A.; *et al.* HUMN project initiative and review of validation, quality control and prospects for further development of automated micronucleus assays using image cytometry systems. *Int. J. Hyg. Environ. Health* **2013**, *216*, 541–552. [CrossRef] [PubMed]

20. Norppa, H.; Bonassi, S.; Hansteen, I.L.; Hagmar, L.; Strömberg, U.; Rössner, P.; Boffetta, P.; Lindholm, C.; Gundy, S.; Lazutka, J.; *et al.* Chromosomal aberrations and SCEs as biomarkers of cancer risk. *Mutat. Res.* **2006**, *600*, 37–45. [CrossRef] [PubMed]

21. Bonassi, S.; Norppa, H.; Ceppi, M.; Strömberg, U.; Vermeulen, R.; Znaor, A.; Cebulska-Wasilewska, A.; Fabianova, E.; Fucic, A.; Gundy, S.; *et al.* Chromosomal aberration frequency in lymphocytes predicts the risk of cancer: Results from a pooled cohort study of 22,358 subjects in 11 countries. *Carcinogenesis* **2008**, *29*, 1178–1183. [CrossRef] [PubMed]

22. Fenech, M.; Kirsch-Volders, M.; Natarajan, A.T.; Surralles, J.; Crott, J.W.; Parry, J.; Norppa, H.; Eastmond, D.A.; Tucker, J.D.; Thomas, P. Molecular mechanisms of micronucleus, nucleoplasmic bridge and nuclear bud formation in mammalian and human cells. *Mutagenesis* **2011**, *26*, 125–132. [CrossRef] [PubMed]

23. Kirsch-Volders, M.; Bonassi, S.; Knasmueller, S.; Holland, N.; Bolognesi, C.; Fenech, M.F. Commentary: Critical questions, misconceptions and a road map for improving the use of the lymphocyte cytokinesis-block micronucleus assay for *in vivo* biomonitoring of human exposure to genotoxic chemicals-A HUMN project perspective. *Mutat. Res.* **2014**, *759*, 49–58. [CrossRef] [PubMed]

24. Azqueta, A.; Collins, A.R. The essential comet assay: A comprehensive guide to measuring DNA damage and repair. *Arch. Toxicol.* **2013**, *87*, 949–968. [CrossRef] [PubMed]

25. Torres-Bugarín, O.; Zavala-Cerna, M.G.; Nava, A.; Flores-García, A.; Ramos-Ibarra, M.L. Potential Uses, Limitations, and Basic Procedures of Micronuclei and Nuclear Abnormalities in Buccal Cells. *Dis. Markers* **2014**, *2014*. [CrossRef] [PubMed]

26. Rojas, E.; Lorenzo, Y.; Haug, K.; Nicolaissen, B.; Valverde, M. Epithelial cells as alternative human biomatrices for comet assay. *Front. Genet.* **2014**, *5*. [CrossRef] [PubMed]

27. Sprent, J.; Tough, D.G. Turnover of native and memry phenotype T cells. *J. Exp. Med.* **1994**, *179*, 1127–1135.

28. Dusinska, M.; Collins, A.R. The comet assay in human biomonitoring: Gene-environment interactions. *Mutagenesis* **2008**, *23*, 191–205. [CrossRef] [PubMed]

29. Anderson, D.; Dhawan, A.; Laubenthal, J. The comet assay in human biomonitoring. *Methods Mol. Biol.* **2013**, *1044*, 347–362. [PubMed]

30. Zock, J.P.; Rodriguez-Trigo, G.; Pozo-Rodriguez, F.; Barberà, J.A.; Bouso, L.; Torralba, Y.; Antó, J.M.; Gómez, F.P.; Fuster, C.; Verea, H.; *et al.* Prolonged respiratory symptoms in clean-up workers of the *Prestige* oil spill. *Am. J. Respir. Crit. Care Med.* **2007**, *176*, 610–616. [CrossRef] [PubMed]

31. Fenech, M. Cytokinesis-block micronucleus cytome assay. *Nat. Protoc.* **2007**, *2*, 1084–1104. [CrossRef] [PubMed]

32. ISCN. *An International System for Human Cytogenetic Nomenclature*; Shafer, L.G., McGowan-Jordan, J., Schmid, M., Eds.; S. Karger: Basel, Switzerland, 2013.

33. Liang, K.Y.; Zeger, S.L. Longitudinal data analysis using generalized linear models. *Biometrika* **1986**, *73*, 13–22. [CrossRef]

34. De la Chica, R.A.; Ribas, I.; Giraldo, J.; Egozcue, J.; Fuster, C. Chromosomal instability in amniocytes from fetuses of mothers who smoke. *JAMA* **2005**, *293*, 1212–1222. [CrossRef] [PubMed]

35. Thompson, S.L.; Compton, D.A. Chromosomes and cancer cells. *Chromosome Res.* **2011**, *19*, 433–444. [CrossRef] [PubMed]

36. Luzhna, L.; Kathiria, P.; Kovalchuk, O. Micronuclei in genotoxicity assessment: From genetics to epigenetics and beyond. *Front. Genet.* **2013**, *4*, 131. [CrossRef] [PubMed]

37. Ishioka, N.; Umeki, S.; Hirai, Y.; Akiyama, M.; Kodama, T.; Ohama, K.; Kyoizumi, S. Stimulated rapid expression *in vitro* for early detection of *in vivo* T-cell receptor mutations induced by radiation exposure. *Mutat. Res.* **1997**, *390*, 269–282. [CrossRef]

38. Littlefield, L.G.; Joiner, E.E. Analysis of chromosome aberrations in lymphocytes of long-term heavy smokers. *Mutat. Res.* **1986**, *170*, 145–150. [CrossRef]

Journal of
Marine Science and Engineering

MDPI

Article

An Oil Fate Model for Shallow-Waters

Juan M. Restrepo [1,*], **Jorge M. Ramírez** [2,†] **and Shankar Venkataramani** [3,†]

[1]   Department of Mathematics, Oregon State University, Corvallis, OR 97331, USA
[2]   Departamento de Matemáticas, Universidad Nacional de Colombia, Sede Medellín, Medellín Colombia;
      E-Mail: jmramirezo@unal.edu.co
[3]   Mathematics Department, University of Arizona, Tucson, AZ 85721, USA;
      E-Mail: shankar@math.arizona.edu
*   Author to whom correspondence should be addressed; E-Mail: restrepo@math.oregonstate.edu;
      Tel.: +1-520-990-4866.
†   These authors contributed equally to this work.

Academic Editor: Merv Fingas
Received: 15 October 2015 / Accepted: 26 November 2015 / Published: 4 December 2015

Abstract: We introduce a model for the dynamics of oil in suspension, appropriate for shallow waters, including the nearshore environment. This model is capable of oil mass conservation and does so by evolving the oil on the sea surface as well as the oil in the subsurface. The shallower portion of the continental shelf poses compounding unique modeling challenges. Many of these relate to the complex nature of advection and dispersion of oil in an environment in which wind, waves, as well as currents all play a role, as does the complex bathymetry and the nearshore geography. In this study we present an overview of the model as well as derive the most fundamental of processes, namely, the shallow water advectiion and dispersion processes. With regard to this basic transport, we superate several fundamental challenges associated with creating a transport model for oil and other buoyant pollutants, capable of capturing the dynamics at the large spatio-temporal scales demanded by environmental and hazard mitigation studies. Some of the strategies are related to dimension reduction and upscaling, and leave discussion of these to companion papers. Here we focus on wave-filtering, ensemble and depth-averaging. Integral to the model is the proposal of an ocean dynamics model that is consistent with the transport. This ocean dynamics model is detailed here. The ocean/oil transport model is applied to a couple of physically-inspired oil-spill problems in demonstrate its specialized capabilities.

Keywords: ocean pollution; oil transport; oil slick; nearshore; ocean oil fate model; oil advection; oil dispersion

---

1. Introduction

The 2010 Gulf of Mexico oil spill, precipitated by a malfunction and an ensuing explosion in the Deepwater Horizon platform prompted several ocean/pollution modeling and simulation groups to exercise their computational platform capabilities. Simulations that attempted to reproduce the long-time fate of oil coming from the Deep-Water Horizon accident yielded very tentative outcomes (see [1–3], and references contained therein). In addition to incomplete complex multi-physics (an atmospheric boundary layer, surface oil, oil at depth, ocean and nearshore dynamics, biogenic dynamics), a further challenge was the sheer spatio-temporal bandwidth required to produce simulations of adequate resolution; from meters to hundreds of kilometers, from tens of seconds, to seasons and years. A very significant complicating fact has been that data that could be used to constrain/tune the model simulations has not been available; particularly, sub-surface oil.

Operational ocean oil evolution models are presently under development. Roughly speaking, there are three types of models in the works: particle-based Lagrangian models [4], mass conservation models [5,6]; there are also models that are specialized to the very difficult task of capturing oil plumes in the neighborhood of underwater spills (see [7] for example). Here we report on the development of a mass conservation oil model, developed specifically for shallow water conditions. Specifically, a model for ocean oil in waters typified by the shelf of the Gulf of Mexico (which is the region in which over ten thousand oil wells presently operate). Hence, we are considering waters of maximal depth of a couple of kilometers and minimal depths in the order of a couple of meters; horizontal scales as large as the Gulf and as small as tens of meters; time scales in the order of 10 s up to months.

A next generation hydrocarbon fate model is highly desirable; one that is capable of simulating accurately oil spills of the magnitude and complexity of the Deep-Water Horizon event. This would be a model with a vast spatio-temporal range: capable of resolving horizontal scales in the order of tens of meters to hundreds of kilometers, and time scales from tens of seconds to seasons. Present-day surface oil models capture oil spills that respond overwhelmingly to currents and wind. Their forecasting fidelity, however, fades quickly in time [1,8]. To a large extent the specific ocean circulation that couples to the oil transport equations has a dramatic effect on the fate of oil and pollutants. However, even if the ocean component of the model is improved, present day oil-fate models have many critical shortfalls. For example, (1) many oil models do not conserve oil mass (models that can faithfully account for subsurface and surface oil over large spatio-temporal scales). Accounting for surface and subsurface oil through data is already extremely challenging, a good model that can be used to explore scenarios would be very helpful; (2) oil can diffuse to and/or agglomerate at subgrid scales (oil is mostly a collection of drops), particularly in rough oceans and hence the dispersion model in the transport equations has to be able to account for this. Both of these effects create very large uncertainties in mass conservation in oil models make it impractical to study how dispersants could change the course of a disaster: how it should be applied, where, and how much; (3) oil is made up of thousands of chemicals. Oil models that lump these chemicals into just a few species create chemical composites with resulting chemistry that is often captured by reactions with many tunable parameters parameters that are hard to constrain via field data.

We introduce a new oil-fate model in this study. The long term goal is to produce a circulation model that captures oil spill dynamics in the shallower regions of the continental shelf, and the nearshore, at the very large spatio-temporal scales required of environmental studies. In order to reach these enormous scales a combination of filtering and of upscaling is required. Because of this the transport model is relatively distinguished. The waves, currents, and atmospheric dynamics would be provided by existing, well-maintained circulation models. The engineering of the interface between the transport model and these circulation models has to make it possible for the oil-fate platform to reap of improvements on ocean/atmosphere dynamics without minimal impact on the code for the oil transport. Moreover, the interface must also provide directives and facilities for the upscaling and filtering, leading to consistent spatio-temporal scales between the oil and the ocean/atmosphere physics. A schematic of the eventual computation platform appears in Figures 1–3.

Tackling nearshore/shelf complexities extends beyond the Gulf region: similar scales and phenomenology exist in other parts of the World, (e.g., the Middle Atlantic Bight, the nearshore of the Great Lakes, the Mediterranean, the Caspian Sea). In these regions bathymetric effects are crucial [9,10], mixing/turbulence, the interaction of the buoyant oil with the sediment [11,12], and sources of freshwater important factors [13–16]. We are thus formulating a tool for critical decision-making of wide applicability. Intended for oil dynamics, some of the results and modeling strategies should extend to other pollutants of interest in hazard prediction and abatement analyses.

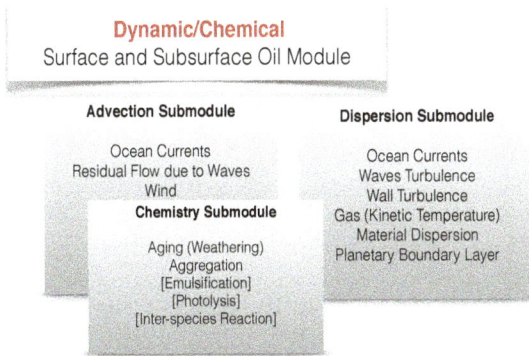

Figure 1. Advection and dispersion aspects of the model that are implemented in these modules are examined in detail in this study. Brackets indicate that a suitable empirical representation of the phenomenon would be used to capture these in the fully developed model. The ocean, atmosphere, and waves are captured by existing, supported circulation models. See Figures 2 and 3.

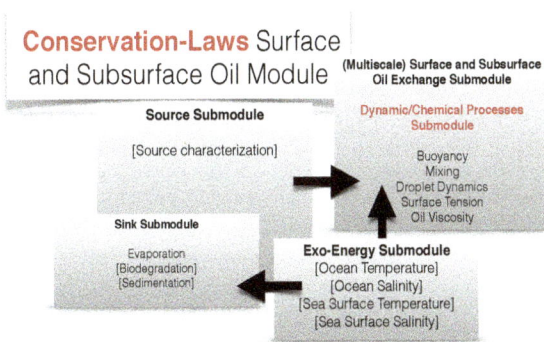

Figure 2. Mass and Energy Conservation Module used to evolve the dynamics of surface and subsurface oil and their exchanges. The multiscale exchange module distinguishes this model from other oil transport models. See Figures 1 and 3.

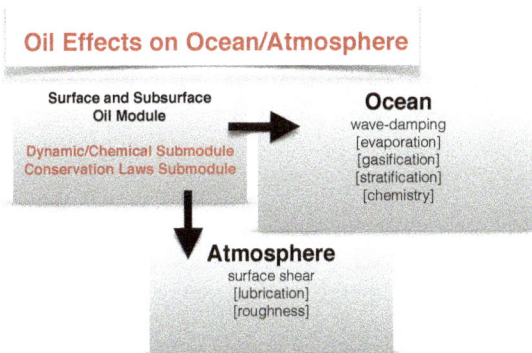

Figure 3. Effects of oil on the atmosphere and the ocean that will be included in the model development. See Figures 1 and 2.

J. Mar. Sci. Eng. **2015**, *3*, 1504–1543

Without being exhaustive we mention in Section 2 several existing oil-fate models. With this context it is easier to appreciate in what respects our model is similar as well as different from others. Some of these models have many years of development and testing, but they are all works in progress. The overall trend in their development is some degree of specialization with regard to scope of applicability. Our oil fate model will track the evolution of surface and subsurface oil. The evolution equations are described in Section 3. In order to achieve large spatio-temporal scales required from a model that would be useful in environmental studies we will utilize filtering and upscaling. In doing so we forego aspects of the dynamics of the oil: in particular we will not resolve the dynamics of oil at sub-wave scales, nor will we be able to describe the vertical distribution of oil. The basic premise is that capturing the total mass of submerged and surface oil over vast regions is more important than resolving details of the oil distribution inside the water column. In this study we will focus on the advective and dispersive aspects of the model. Specifically, we will highlight the interaction of the wind, waves, and currents, with oil in suspension in the upper turbulent mixing layer, and the oil slick itself. There are three other processes that are critical to formulating a practical and reasonably complete computational platform that simulates the oil model. The most critical of these, for mass conservation is the model for the mass exchanges between the surface and the sub-surface oil components. In this study we briefly describe in Section 3.4.1 the underlying model briefly, leaving concrete details for another paper. Another critical component is phenomena associated with chemistry. We have opted to adopt the models for the chemistry developed by groups who specialize in this aspect of oil in ocean environments. Instead we have focused our attention on a practical problem related to the eventual implementation of the model in the form of a computational platform. Capturing weathering or aging of oil. By aging we mean specifically to the resulting complex time-dependent reactive or dissipative aspects of simulating a chemically-reacting liquid which is composed of thousands of basic chemicals. In a separate study we describe how an upscaling strategy that we propose can lead to a significant computational gain by producing a dimension reduction while at the same time circumventing the inherent stiffness of reactive/evaporative processes in a model that also has to capture advection and diffusion. Aging is discussed in general terms in Section 3.4.2. Section 3 concludes with a summary of other phenomena that will be incorporated in the oil-fate model. Among these are, emulsification, photolysis, sedimentation. Along with oil source modeling, these phenomena are studied in depth by other researchers and we intend to incorporate into our model their findings.

The upscaling and filtering extends to the ocean dynamics, and in Section 4 and Section 5 we describe the ocean dynamics and the energy, scale-compatible with the oil transport model. Of note is the importance given to wind and wave effects on the advection and the dispersion of the transport equation. In Section 6 we describe two applications of the model in which we illustrate the important role played by waves in the evolution of oil in the nearshore and on the shelf. A reprise of the model and of the results of the model illustrations appears in Section 7.

2. Background

There are a variety of commercial and non-commercial oil spill simulation platforms. OILMAP/ASA Sciences is a general model and it is often used by large oil companies and by government agencies of over 40 countries in their oil hazards planning. This model achieves generality and computing capabilities at the expense of accuracy in physics and detail. It works well in deep water spills involving very light crude. Among the specialized oil spill modeling efforts, we can mention OSCAR/SINTEF. SINTEF is the commercial developer of OSCAR. The platform has been used for environmental decision-making and forecasting in the Northern Sea, the Gulf, the Mediterranean, among other places. Within the SINTEF development effort one can find more specialized models, such as OWM (Oil Weathering Model, similar to ADIOS), DeepBlow (blowouts, deep sea drilling), DREAM and ELMO (risk analysis, dispersants), Partrack (drill muds and cuttings). SINTEF also develops SINMOD, their primitive equation circulation, which has no wave effects and is thus of limited use in nearshore environments. The Mediterranean Decision Support System for Marine Safety

(MEDESS-4EMS) is a European Union consortium that includes a variety of different oil modeling platforms, such as MOTHY, POSEIDON-OSM, MEDSLICK. Each of these models have different capabilities and uses and have predictive capabilities in the short term and far from the shores. Among the models with shelf capabilities, there is COZOIL [17] and VOILS [7], the latter is designed to handle estuarine environments and thus will be used in the future in our intermodel comparisons. Among the best with regard to the physics we mention VDROP and VDROP-J [18]. These last ones are primarily intended to model underwater oil plumes, in situations where baroclinicity is not crucial.

3. Oil Dynamics

A schematic of the domain, along with the coordinate system is described in Figure 4. Time is denoted by $t, T \geq 0$, where the former is changing at wave scales and the latter at the longer scales of interest. The transverse coordinates of the domain will be denoted by $\mathbf{x} := (x, y)$. The cross-shore coordinate is x and increases away from the beach. The vertical coordinate is z, with $z = 0$ corresponding to a quiescent ocean. The sea surface is described by $z = \eta(\mathbf{x}, t)$, and relative to η, we posit the existence of a very thin layer of oil with thickness $S(\mathbf{x}, t)$. This is the oil slick. Substantial amounts of oil may be present in the turbulent mixed layer neighboring the ocean surface, and often in a fine mist, in the bulk of the water column; formally, the thickness of S is determined by whether it is mostly composed of oil. The bottom of the ocean is described by $z = -h(\mathbf{x})$, fixed in time. The total water column depth is given by $H(\mathbf{x}, t) := h(\mathbf{x}) + \eta(\mathbf{x}, t) + S(\mathbf{x}, t) \approx h(\mathbf{x}) + \eta(\mathbf{x}, t)$. Spatial differential operators may be split into their transverse and vertical components, *i.e.*, $\nabla = (\nabla_\perp, \partial_z)$, where the first entry depends only on $\mathbf{x}$. A schematic of the domain, along with the coordinate system is described in Figure 4.

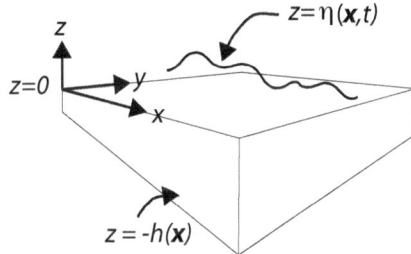

Figure 4. Schematic of the nearshore environment, z increases above the quiescent level of the sea $z = 0$, $\mathbf{x} := (x, y)$, and t is the time variable. The sea elevation $z = \eta$, includes a component of the free surface associated with the currents is $z = \eta(\mathbf{x}, t)$ and a component associated with waves. The bottom topography $z = -h(\mathbf{x})$ is referenced to the quiescent sea level height, $z = 0$.

The sea elevation is further split $\eta = \zeta^c + \eta^w$, where ζ^c is the component that changes at scales much greater than wave scales, and η^w is the component associated with waves.

3.1. The Oil Slick Component

The aim is to propose a model for the evolution and fate of ocean oil, consistent with the spatio-temporal scales of interacting waves and currents in coastal waters. Specifically, with the wave/current interaction model proposed in [19]. A full accounting of the [19] wave/current interaction model and its asymptotic balances will not be repeated here.

We suppose that oil at the surface of the ocean is composed entirely of a multi-species incompressible fluid hydrocarbon mixture. We denote this as the oil slick. Its total mass is

$$M_s(t) = \int_{\Omega_t} \sum_{i=1}^{N} \tilde{s}_i(\mathbf{x}, t) \rho_i dx$$

where i distinguishes the various chemical components that make the oil slick. Typical crude is a complex combination of chemicals. As reported in [20], oil modelers will define the chemical components to track by lumping types (e.g., alkenes, alkanes, aromatics, *etc.*), or by lumping individual components by their similarity in boiling point. The latter is usually a practical way to divide an oil spill into subcomponents because the boiling point of the chemicals making up an oil spill will be known because they are necessary in the refining process.

The mass will depend on time if the spatial domain Ω_t in question depends on time, e.g., if oil is allowed to flood dry land, in which case the basin is allowed to change in time. Each chemical component has a known density ρ_i. The total depth of the oil slick is

$$S = \sum_{i=1}^{N} \tilde{s}_i$$

It is also assumed that there is oil in the bulk of the water column. We will denote this oil, the *subsurface oil*.

We assume that $|S| \ll 1$, so that the dynamic pressure drop in the thin layer is $p = \mathcal{O}(|S|)$. Curvature effects are ignored, *i.e.* the outward normal to the ocean surface

$$\hat{\mathbf{n}} = \frac{\nabla g(\mathbf{x}, z, t)}{|\nabla g(\mathbf{x}, z, t)|} \approx \hat{\mathbf{z}}, \quad \text{where } g(\mathbf{x}, z, t) = z - \eta(\mathbf{x}, t) - S(\mathbf{x}, t) = 0$$

and thus continuity of stresses at the free surface simplify:

$$\mu_i \frac{\partial \tilde{u}_i}{\partial z} = \tau(\mathbf{x}, t) \tag{1}$$

where $\tilde{u}_i$ is the velocity in the film, μ_i is the oil viscosity, and τ is the transverse component of the wind stress. The film is thin and de-void of an inflection point. So in its most general form it is a quadratic function of the layer thickness. The velocity condition at the oil/water interface $z = \eta$, is

$$\tilde{u}_i = C_{xs} \mathbf{u} \tag{2}$$

where $\mathbf{u}$ is the Eulerian ocean velocity, evaluated at the ocean surface. The parameter $C_{xs} \geq 0$ is a modeling parameter that accounts for large-scale manifestations of oil droplet inertia and cross section.

The momentum equations in the oil slick are

$$\rho_i \tilde{u}_i \cdot \nabla \tilde{u}_i = -\nabla \tilde{p}' + \mu_i \Delta \tilde{u}_i$$

where we have ignored the fast time changes in $\tilde{u}_i$, *i.e.*, we make an adiabatic assumption. Since the pressure $\tilde{p}' = p_0 + \tilde{p}$, where p_0 is the ambient pressure and $\tilde{p}$ is the dynamic pressure not dependent on z. We assume that $\partial_x^\eta \ll \partial_z^\eta$, and $|\nabla_\perp \eta| < 1$. Let

$$\nabla \Pi := \nabla \tilde{p}' + \rho_i \tilde{u}_i \cdot \nabla \tilde{u}_i = \nabla p' + \rho_i \frac{1}{2} \nabla |\tilde{u}_i|^2$$

(The vertical component of the velocity in the oil slick is approximated by the vertical component of the ocean velocity). Hence,

$$\nabla \Pi = \mu_i \frac{\partial^2 \tilde{u}_i}{\partial z^2}$$

Integrating in z twice, applying the boundary conditions (Equations (2) and (1)) we obtain an expression for the transverse velocity

$$\tilde{u}_i = \frac{1}{2\mu_i}(z-\eta)^2 \nabla \Pi + \frac{1}{\mu_i}\left[\tau - s_i \nabla \Pi\right](z-\eta) + C_{xs}\mathbf{u}$$

The pressure gradient is a response to the oil slick curvature, which at these scales approximates to

$$\nabla \Pi \approx -\gamma_i^t \nabla \cdot \nabla s_i \tag{3}$$

where γ_i^t is the surface tension constant associated with species i. We define a depth-averaged velocity

$$\check{u}_i = \frac{1}{s_i}\int_\eta^{\eta+s_i} \tilde{u}_i dz \tag{4}$$

Wave averaging is presumed and the time scale $T \gg 2\pi/\sigma_0$, the time scale for the waves of dominant frequency σ_0. In the above equation we have omitted losses/gains in oil, for simplicity, but these effects will be added later.

Using this expression for the velocity in the layer, substituting (Equations (3)), and performing the average (Equations (4)), we obtain the flux

$$\check{u}_i s_i = C_{xs}s_i\mathbf{u} + \frac{1}{2\mu_i}\tau s_i^2 + \frac{\gamma_i^t}{3\mu_i}\Delta s_i s_i^3$$

We are interested in developing dynamic equations at scales that are large compared to those typical of waves. We will define the *wave*-average of some quantity f, as

$$\bar{f}(\cdot, T) = \frac{2\pi}{\sigma_0}\int_T^{T+\sigma_0/2\pi} f(\cdot, t)\, dt \tag{5}$$

where σ_0 is the dominant wave frequency. For monochromatic waves, this averaging is the same as Reynolds averaging. For more complex spectra *and* wave climatology the wave average and the Reynolds average are not the same, though we will be using the two terminologies interchangeably in what follows.

Using the continuity equation and depth-averaging, retaining up $\mathcal{O}(s_i^2)$ we obtain the oil slick equation:

$$\frac{\partial s_i}{\partial T} + \nabla_\perp \cdot \left(C_{xs}s_i\mathbf{V} + \frac{1}{2\mu_i}\tau s_i^2\right) = \dots$$

where the advection velocity at the surface is approximated as

$$\mathbf{u} \approx \mathbf{V}(\mathbf{x}, T) := \mathbf{v}^c + \mathbf{u}^{St} \tag{6}$$

$\mathbf{V}$ is the depth-averaged transport velocity (*cf.* [21] and [19]). It has contributions from the depth-averaged current $\mathbf{v}^c$ and the Stokes drift velocity $\mathbf{u}^{St}$ as well as the residual velocity due to wave breaking (see [22]). As will be discussed in Section 4, $\mathbf{V}$ is supplied by the wave/current interaction equations. Approximating the surface velocity $\mathbf{u}$ by $\mathbf{V}$ is predicated on the fact that the transverse velocity is qualitatively similar, at the large spatio-temporal scales we have in mind. The appearance of the Stokes drift velocity, which can be comparable to the Eulerian ocean current in the nearshore, antecedes the inclusion of dispersive corrections to the oil slick transport equation and is elaborated upon in Section 3.2. The wind stress provides more local corrections via the second term in the advection term. The wind stress related advection can dominate the ocean related advection if $s_i|\boldsymbol{\tau}|/2\mu_i > C_{xs}|\mathbf{V}|$. We also note that the variability of the $\boldsymbol{\tau}$ is usually much larger and changes much faster in time than other advective mechanisms.

Oil buoyancy forces cannot be neglected. Crude oil, being a composite of several chemicals, has aggregate complex buoyancy characteristics. The composition of the oil in the oil slick changes over time because different species of chemicals evolve differently. Moreover, some of these species react chemically in the presence of sun light. It is thus essential to track each species separately: Each species can have a different evaporation rate b_i. Reactions due to photosensitivity can be handled adiabatically since these reactions among the species occur at time scales much shorter than the time scales of the oil slick dynamics.

Hence, for the ith species, the equation is

$$\frac{\partial s_i}{\partial T} + \nabla_\perp \cdot \left(C_{xs}\mathbf{V}s_i + \frac{\boldsymbol{\tau}}{2\mu_i}s_i^2 \right) = \nabla_\perp [\Psi\nabla_\perp s_i] - E_i^s(s_i) + R_i + P_i^s + G_i^s \tag{7}$$

The first term on the right hand side is the eddy/dispersion term and will be described in detail when we discuss sub-surface oil, in Section 3.2. The *reaction* term $E_i^s(b_i, s_i, s_j)$ encompasses chemical reactions among species within the slick as well as other processes modeled by rates, such as evaporation. R_i is the rate associated with mass exchanges between the oil slick and the interior of the ocean (See Section 3.4.1). P_i^s is the rate of biodegradation, emulsification, photodegradation and G_i^s is the source rate term. Unlike R_i the terms with s superscript are particular to the surface oil. All of the terms on the right hand side have units of mass per unit time per unit area.

The transport equation for the oil slick, (Equation (7)) along with its initial condition and boundary conditions, is coupled to a model for subsurface oil and ocean, which we discuss next.

3.2. The Sub-Surface Oil Component

The heavier oil will sediment out quickly and the rest is found in suspension. Most of this oil, being buoyant, is found near the ocean surface at least away from sources and sinks.

The total mass of the oil in the sub-surface is defined as

$$M_c = \int_{\Omega_t} \int_{-h}^{\eta} \sum_{i=1}^{N} C_i(\mathbf{x}, z, T)\, dz\, d\mathbf{x}$$

where C_i is the concentration of species i in the mixed layer. These concentrations have dimensions of density. The index i identifies the union of similar species in the surface and subsurface layers. Our goal is to produce a simple model for depth-averaged sub-surface oil. While doing so leads to a computationally more efficient model, it forgoes depth-dependence details of the subsurface oil. Presumably these details are less critical on vast expanses in shallow waters.

In what follows we will omit the i subscript, as we will be referring to a single species of tracer with concentration $B(\mathbf{x}, z, t, T) = \overline{B}(\mathbf{x}, z, T) + B'$ with B' having zero wave mean. Similarly, let the subsurface ocean transport velocity be written in terms of a Reynolds decomposition as $\mathbf{U} = (U, V, W) = \overline{\mathbf{U}} + \mathbf{U}'$, with $\overline{\mathbf{U}'} = 0$, which includes the current as well as the Stokes drift velocities. The starting point is the Equation (32) from [21]:

$$\frac{\partial B}{\partial t} + \mathbf{U} \cdot \nabla B - \kappa\nabla \cdot [\nabla B] = 0 \tag{8}$$

where κ is the molecular diffusion. Averaging Equation (8) via Equation (5), we obtain

$$\frac{\partial \overline{B}}{\partial T} + \overline{\mathbf{U}} \cdot \nabla\overline{B} - \nabla \cdot [\Psi\nabla\overline{B}] = 0 \tag{9}$$

The term Ψ subsumes the aspect related to molecular diffusion as well as the eddy fluxes, $\mathbf{F}_{i,i} = \overline{U'B'}_{i,i}$. The deviation B' represents a departure from when the concentration is equal to its mean, then $B' \approx -\mathbf{Z} \cdot \nabla\overline{B}$, where $\mathbf{Z} = (Z_1, Z_2, Z_3)$ is the parcel coordinate, assuming that changes in $|\mathbf{Z}|$ are much smaller than changes of $|\nabla\overline{B}|$. $\mathbf{Z}$ satisfies the equation

$$\frac{\partial \mathbf{Z}}{\partial t} + \mathbf{U} \cdot \nabla \mathbf{Z} - \kappa \nabla^2 \mathbf{Z} = \mathbf{U}'$$

Returning to the derivation of the large-scale transport equation, we consider the tensor

$$\Psi_{ij} = \kappa \delta_{ij} + \overline{U_i' Z_j}$$

where $i, j = 1, 2, 3$, δ is the Kronecker delta. The tensor

$$\Psi_{ij} := \Psi_{ij}^s + \Psi_{ij}^a$$

as shown in [23], can be written in terms of its symmetric and asymmetric parts. The symmetric tensor Ψ^s is, in principle, diagonalizable. Its effect on $\overline{c}$ are advective and parallel to the contours of $\overline{B}$:

$$2\Psi_{ij}^s = 2\kappa \overline{Z_{i,l} Z_{j,l}} + \overline{Z_i Z_j}_{,t} + \{\overline{U_l Z_i Z_j} - \kappa \overline{Z_i Z_j}_{,l}\}_{,l}$$

(The comma anteceding an index in the notation ",l" is interpreted as the derivative with respect to the lth coordinate, $l = 1, 2, 3$, and repeated indices imply summation). Although κ is usually very small compared to the eddy viscosity it guarantees Ψ^s positive-definiteness.

The antisymmetric part of the eddy flux is associated with a diffusion effect in directions not necessarily orthogonal to contours of the wave-averaged tracer unless Ψ^a is isotropic.

$$2\Psi_{ij}^a = -2\kappa \overline{Z_{i,l} Z_j - Z_{j,l} Z_i}_{\,l} - \overline{Z_i Z_{j,t}} - \overline{Z_j Z_{i,t}} + \overline{Z_j U_k Z_{i,l}} - \overline{Z_i U_k Z_{j,l}}$$

Translating [23] to our own notation, the asymmetric (or skew) part of the flux

$$F_{i,i}^a = -[\Psi_{ij}^a \overline{B}_{,j}]_{,i} = [\epsilon_{ijl} G_l \overline{B}_{,j}]_{,i} = (\mathbf{G} \times \nabla\overline{B})_{i,i}$$

where $\mathbf{G} := (D_{23}^a, D_{31}^a, D_{12}^a)$, and ϵ_{ijl} is the cyclic operator. The divergence of the skew flux is then

$$\nabla \cdot \overline{U'B'}^a = -\nabla \times \mathbf{G} \cdot \nabla\overline{B}$$

The contribution of the antisymmetric component is advection of the mean tracer by the Stokes drift $\mathbf{u}^{St} = \nabla \times \mathbf{G}$, and has already been accounted for in Equation (9), hence care needs to be exercised so that it is not double counted in the advection.

We proceed by depth-averaging Equation (9). With $H \equiv h + \eta$, the depth-average of f, say,

$$\hat{f}(\mathbf{x}, T) = \frac{1}{H} \int_{-h}^{\eta} f(\mathbf{x}, z, T) dz$$

The depth and wave averaged oil concentration is defined as

$$C(\mathbf{x}, T) := \widehat{\overline{B}}(\mathbf{x}, z, T) \qquad (10)$$

while the corresponding velocity (see Equation (6)) is

$$\mathbf{V}(\mathbf{x}, T) := \widehat{\overline{\mathbf{U}}} = \mathbf{v}^c + \mathbf{u}^{St} \qquad (11)$$

The unfortunate characteristic about depth-averaging dynamic variables is that, unlike wave averaging, the fluctuating component does not have zero mean, and further, wave averaging and depth averaging are not commuting operations. We hope, however, that the generally fluctuating component and its derivative of either averaging remain small and/or that the correlations between the averaged quantity and the fluctuating field are small.

The first term in Equation (9), when integrated with respect to z yields

$$\frac{\partial}{\partial T} \int_{-h}^{\eta} \overline{B} dz - (\overline{B}|_{z=\eta}) \frac{\partial \eta}{\partial T}$$

Since $\nabla \cdot \overline{\mathbf{U}} = 0$, the second term of Equation (9) becomes

$$\nabla_{\perp} \cdot \int_{h}^{\eta} \overline{\mathbf{U}} \, \overline{B} dz - \nabla_{\perp} \eta \cdot (\overline{\mathbf{U}} \, \overline{B}|_{z=\eta}) - \nabla_{\perp} h \cdot (\overline{\mathbf{U}} \, \overline{B}|_{z=-h}) + (\overline{W} \, \overline{B}|_{z=\eta}) - (\overline{W} \, \overline{B}|_{z=-h})$$

We note that $\eta_T + \mathbf{U} \cdot \nabla \eta = W$ at $z = \eta$ and that $W = -\nabla h \cdot \mathbf{U}$ at $z = -h$, so combining the above two expressions yields:

$$\frac{\partial \overline{B}}{\partial T} + \overline{\mathbf{U}} \cdot \nabla \overline{B} = \frac{\partial}{\partial T} \int_{-h}^{\eta} \overline{B} dz + \nabla_{\perp} \cdot \int_{-h}^{\eta} \overline{\mathbf{U}} \, \overline{B} dz$$

Finally, we integrate the third term in (9) to obtain

$$-\nabla_{\perp} \cdot \int_{-h}^{\eta} (\Psi \nabla \overline{B}) dz + \text{averaged fluxes at top and bottom}$$

Adding these and depth averaging,

$$\frac{\partial HC}{\partial T} + \nabla_{\perp} \cdot (H\mathbf{V}C) = \nabla_{\perp} \cdot [H\Psi \nabla_{\perp} C] + \rho \left[E^C(C) - R + P^C + G^C \right] \tag{12}$$

On the right hand side, the first term is the resulting dispersion in units of length-squared over time, which captures the familiar Reynolds stresses contributions as well as dispersion due to the depth averaging. The dispersion is usually parametrized, and we shall denote this parametrization of Ψ by D. The reaction/chemistry term is E^C, the mass exchange term R. P^C represents the sedimentation and biodegradation mechanisms. The subsurface source term is G^C. These last four terms have dimensions of mass per unit time per unit area.

In shallow waters and far from sources and sinks, the largest concentration of oil, particularly if it is light crude, will be in waters close to the surface of the ocean, the mixed layer. In the above we are averaging over the total water column depth, however, dispersion in the nearshore is expected to depend highly in the topography as well as the roughness of the wavy surface. Clearly, close to the ocean surface there is a highly turbulent breaker layer with thickness $h_b \propto a_*$, where a_* is the typical size of the waves, which is in the order of a meter (see [22]). In the deeper reaches of the shelf, and at very large spatio-temporal horizontal scales, we expect an Ekman layer $h_{Ek} \propto v_*/f$, which can be as large as 100–200 meters. f is the Coriolis parameter. However, for most of the shelf we expect that mixing is dominated by Langmuir turbulence. At intermediate times, t', between the wave and the current scales, the adjustment concentration b due to wave dispersion (see Eq. 10.7 in MRL04) is obtained by solving

$$\frac{\partial b}{\partial t'} - \frac{\partial}{\partial z} \left[\frac{\partial \frac{1}{2} e^2}{\partial t'} \frac{\partial}{\partial z} b \right] \approx 0, \tag{13}$$

Here e^2 is the wave variance, which to to first order is

$$\frac{1}{2} \frac{\partial e^2}{\partial t'}(\mathbf{x}, T) \approx \frac{1}{2} \frac{A^2}{g} \frac{\sinh[k(z+H)]}{\sinh[kH]}.$$

where σ is the wave angular frequency, k is the magnitude of the wave number, and A is the wave action. (These last quantities are discussed in detail in Section 4). Since we are interested in the dynamics of oil at the largest spatio-temporal scales, we subsume the intermediate fluctuations into the larger scale quantity representing the oil dynamics at the scales of interest. There is, however, a useful, but not surprising estimate that can be derived from the steady variant of Equation (13): the distance that defines a Langmuir turbulence thickness, present near the surface of the ocean, is $z/H \sim \min(1, 1/kH)$. For large kH it is the distance that the concentration at the surface drops by e^1 by diffusive processes. We will thus define

$$P := \min(H, 1/k) \tag{14}$$

as the Langmuir mixed layer depth. A measure of the relative importance of the windshear to the shearing within the Stokes drift layer is

$$\mathrm{La} := \sqrt{v_*/u^{st}(0)}$$

Here $u^{st}(0)$ is the Stokes drift velocity evaluated at $z = 0$. On the Gulf Coast Shelf, La is in the order of 1, the typical friction velocity is less than 10 cm/s.

3.3. Dispersion

The tensor Ψ in Equation (12) has to capture a variety of causes for transverse dispersion: the Reynolds stresses that arise from the filtering, as well as the more complicated dispersion fluxes due to depth averaging. We write

$$\Psi = \Sigma + \Xi \approx D \tag{15}$$

where

$$\Sigma_{ij} = \frac{1}{H} \int_{-h}^{\eta} (\overline{U_i'U_j'} - \overline{U}_i\overline{U}_j)dz$$

is the depth averaged transverse turbulent Reynolds stress, and

$$\Xi_{ij} = \frac{1}{H} \int_{-h}^{\eta} [(\overline{U} - V)_i(\overline{U} - V)_j] \, dz - \frac{1}{H} \int_{-h}^{\eta} (\overline{U} - V)_i \, dz \frac{1}{H} \int_{-h}^{\eta} (\overline{U} - V)_j \, dz$$

is the dispersion caused by Reynolds averaging and depth averaging. The indices $i, j = 1, 2$ refer now to transverse coordinates. At this point we need to make a choice for the parametrizations of both of these tensors. For Σ we choose a depth-averaged Smagorinsky type, grid based parametrization and further, that dispersion is orthogonal to the gradient of the tracer. Adopting the suggestion in [24],

$$
\begin{aligned}
\Sigma_{ij} &= \alpha \max(dx, dy)\delta_{ij} \frac{1}{H} \int_{-h}^{\eta} [\overline{U}_{,i}^2 + \overline{U}_{,j}^2 - \frac{1}{2}(\overline{U}_{,i} + \overline{U}_{,j})^2]^{1/2} dz \\
&\approx \alpha' \max(dx, dy)\delta_{ij} [V_{,i}^2 + V_{,j}^2 - \frac{1}{2}(V_{,i} + V_{,j})^2]^{1/2}
\end{aligned} \tag{16}
$$

The dx and dy are the discretization grid spacings and α' is a parameter which needs to be tuned. For the other tensor we propose

$$\Xi_{ij} = \kappa\delta_{ij} + \beta(\Theta H + \Theta^{st}/k)_{ij} \tag{17}$$

where β is another tunable parameter, and

$$\Theta_{ij} := \{|(v_* - v^c)_i||(v_* - v^c)_j|\}^{1/2}$$

is the fluctuation velocity based on the difference between the surface friction velocity $\mathbf{v}_* = \tau_* / \rho$, and the current velocity $\mathbf{v}^c$. Similarly,

$$\Theta_{ij}^{st} := \left\{ |(\mathbf{u}^{st} - \mathbf{u}^{st})_i||(\mathbf{u}^{st} - \mathbf{u}^{st})_j| \right\}^{1/2}$$

is the difference between the surface drift velocity and the depth-averaged counterpart.

3.4. Transformation Mechanisms: Chemistry and Physics of Oil

The three most important transformative mechanisms are the mass exchanges between the slick and the subsurface oil, emulsification, and biodegradation. Photolysis is important in oil at or near the surface and it affects new oil more effectively than old oil. Sedimentation via agglomeration or particulate contamination (e.g., the calcium carbonate rain or interactions with bottom or suspended sediment) is most effective in older oil.

The mass exchanges between the slick and the subsurface are due to wave action, background turbulence, and wind, which will tend to fold in surface oil. At the same time, the oil droplets, particularly those of large enough size, will rise due to buoyancy. Changes in the viscosity and surface tension of oil droplets have a microscale effect that affects the larger scale dynamics that we are interested by altering the mass exchange dynamics. So does emulsification. Emulsification is a material state transition and tracking such a transition is important for its consequences on the dynamics, but also because oil in this material state has different environmental remediation strategies than the fluid-like oil counterpart. The changes over time that oil experiences due to emulsification and chemical reactions is denoted as oil *weathering* or *aging*. Biodegradation see [25] and references contained therein) can be a very impactful transformation mechanism in certain environments. Field data from the Gulf of Mexico oil spill appears to confirm this point.

Some aspects of our oil model that extend beyond advection and diffusion will be adopted from the work of others: specifically, surface evaporation, chemical reactions that affect transport, photolysis, emulsification, sedimentation, and source characterization. Other aspects, on the other hand, will be re-examined and models will be proposed for these. Among those phenomena that will be captured by improved models are the mass exchange and weathering. Detailed development of these will be found in companion studies. The mass exchange is fundamental to our model, and like the weathering problem, is challenging because it involves physics at microscales that we do not want to resolve yet needs to be upscaled in order to be included in the oil transport model.

3.4.1. Mass Exchanges between the Subsurface Oil and the Slick

The microscale physics of droplets plays an essential role in the vertical transport of oil from the ocean surface to the subsurface and vice versa. Oil is essentially a complex conglomerate of oil droplets. The droplets have a wide distribution of sizes and chemical composition. Vertical oil transport is the result of a competition between inertial and drag forces, buoyant and shearing forces, in a complex turbulent fluid flow background. These forces can, in addition to sinking and sending oil aloft, change the oil droplet size distribution as well as the droplet chemistry. The details of the model for the mass exchange term R appear in a companion paper [26]. The elements of the model involve the upscaling of Smoluchowsky-type equations for the distribution of oil droplets (see [27]) and eddies. The droplet-droplet interaction and droplet-eddy interaction are highly dependent on surface tension forces and the oil droplet viscosity. Droplets larger than a critical size become buoyant and rise.

3.4.2. Aging: A Consequence of Grouping Chemicals and Unresolved Physics

Weathering refers to the phenomenon where chemical rates, evaporation or sedimentation rates, *etc.* are time dependent. This might also reflect unresolved physics in the model. Oil is typically composed of tens of thousands chemical species [28]. It is however unlikely that practical application of the model would require tracking more than a few species, or groups of chemicals that encompass chemically similar species, for instance chemical complexes with similar burning temperature or molecular weight. Indeed, evolving the actual amount of chemical species is computationally challenging on large oceanic domains; the system of advection diffusion equations for the species would likely be very ill-conditioned due to the wide range of time scales associated with their reaction rates; most importantly, however, is that we opt for robustness in model outcomes over details.

There is precedent for this type of decomposition in fate models (see [29]). In [30] the oil is divided into groups based upon the boiling point of its subcomponents allowing them to better handle the weathering processes and at the same time obtain a reduction of complexity. When composite species are used, however, one is bound to see *aging* effects. Since some of the chemistry between species is not resolved, the chemical reactions of the subcomponents of oil, each of which is described in simple rate equations, will instead be endowed with complex/time-dependent chemically reacting behavior when agglomerated. The result is complex evaporation and emulsification.

Although the chemistry/evaporation of each compound can be modeled by an autonomous differential equation, this is no longer true for the aggregates. For example, consider a situation where the evaporation rate of the i-th species is well captured by $E_i^C = -b_i C_i$ in Equation (7) with $b_i \geq 0$ a constant rate. Referring now to the i-th aggregate, E_i^C would be a function of rates whose weight will change due to the increase in the relative concentration of the non-volatile components. This phenomenon is called *weathering* and models incorporate this effect by including (often ad hoc or empirical) history dependence in the evolution of the aggregates. We propose an alternative approach using an idea we call *virtual aggregates* which are appropriately chosen linear combinations of the individual species concentrations.

A simple illustration of this idea is as follows: Assume that we have the (autonomous) evaporation equations $\frac{\partial C_i}{\partial T} = -b_i C_i$ for the species concentrations C_i, $i = 1, 2, \ldots, N$, and we only track the aggregate concentration $C = \sum C_i$. If we define the effective evaporation rate $b(T) = -\frac{1}{C} \frac{\partial C}{\partial T}$, then its value is *a priori* uncertain and reflects the uncertainty in the relative concentrations C_i/C of the various species. Absent any further information, a reasonable model for b is an SDE accounting for the uncertainties in the relative concentrations, and with a negative drift that captures the fact that λ is monotonically decreasing in time as the more volatile species evaporate.

We can improve this basic (autonomous SDE) model (or estimate) of b by tracking information that is complementary to the total concentration C. One way to get more information is to track other "moments" of the concentrations $\{C_i\}$ of the form $C_j = \sum \alpha_{j,i} C_i$ for appropriate choices of weights $\alpha_{j,i}$. The quantities C_j are the *virtual aggregates*, and the goal of this type of modeling is to parameterize the (complex, high-dimensional) chemistry of oil by low dimensional, autonomous, stochastic differential equations.

3.4.3. Emulsification and Changes to the Density, Surface Tension, and Viscosity of the Slick

At the spatio-temporal scales our model is destined to operate, viscous effects that affect the dynamics of the slick and the sub-surface oil are overwhelmingly dominated by the eddy viscosity. Nevertheless, it is important to track the evolution of the micro-scale viscosity, as well as the surface tension and density, because of their effect on the balance of forces in the mass exchange term, R in Equation (12). In principle, one would have to track each chemical species' viscosity μ_i. However, the bulk viscosity is not generally a linear combination of the viscosity of different chemicals. In [31], it is suggested that one track the asphaltene content of the complex crude. This defines the "parent oil viscosity" μ_0 estimated to be $\mu_0 \approx 224a^{1/2}$, where a is the percentage of asphaltene in the oil (this empirical relationship is found in [31]).

When emulsification is significant oil turns into a mousse-like substance, mostly from an uptake of water. Some hydrocarbons are not as prone to emulsification (e.g., kerosene, gasoline) which is not

J. Mar. Sci. Eng. **2015**, 3, 1504–1543

the case with oils containing high levels of asphaltenes (see [32]). Taken from [33], the (dimensionless) fractional water content evolves in time as

$$F_{wc}(T) = C_3 \left[1 - \exp\left\{-\alpha_{wc}T\right\}\right] \tag{18}$$

where the rate is

$$\alpha_{wc} = 2 \times 10^{-6}(v_{wind} + 1)^2$$

The dimensionless "mousse viscosity constant" is $C_3 = 0.7$ (for heavy fuel oils and crude and about 0.25 for heating oil, for example), v_{wind} is the wind speed.

The evolution of oil slick viscosity is then computed as a modification on μ_0 using Mooney's Equation (see [34,35]):

$$\mu(T) = \mu_0 (M_e + \exp[C_4 F_{evap}(T)]) \tag{19}$$

with

$$M_e = \exp\left[\frac{2.5 F_{wc}}{1 - C_3 F_{wc}}\right]$$

The second term in Equation (19) accounts to increases in the viscosity due to evaporation where F_{evap} is the fraction evaporated from the slick. F_{wc} is given in Equation (18) and C_4 is a parameter that varies between 1 and 10, the smaller value associated with lighter hydrocarbons.

Emulsification increases the viscosity of oil, but it also increases its solubility and hence its density. The increase in density reads

$$\rho_i(T) = Y(T)\rho_w + (1 - Y(T))(\rho_i(0) + C_3 F_{evap}(T))$$

where ρ_w is the fluid density, and $Y(T)$ is a temperature dependent, nondimensional empirical fit to data. However, observations indicate that de-emulsification also occurs, a process in which water is released. Depending on the chemical composition of the surface oil and the wind speed, the process of emulsification can be stable or meta- or unstable. (See [36] and references contained therein, for more detailed models for emulsification, stability criteria, and original sources of the research on emulsification).

The surface tension increases with evaporation. An empirical formula for the surface tension is

$$\gamma^t(T) = \gamma^t(0)(1 + F_{evap}(T))$$

where $\gamma_s(0)$ is the effective surface tension of the oil slick before being weathered. (See [32] and references contained therein).

Within the oil transport dynamics, the density, viscosity, and surface tension enter the microdynamics of the droplets, which in turn affect mass exchanges between the surface oil component and the sub-surface oil components. The mass exchange rate is R in Equations (7) and (12).

3.4.4. Evaporation

The basic strategy to modeling loses due to evaporation are detailed in [37], and [38]. The model of [39] is widely used (see also [35]). The evaporation rate is

$$E_i^s = -\alpha_i^F K_E^i$$

where K_E^i is an empirical speed (in units m/s), which depends on the local wind speed, the mean diameter of the oil slick, and perhaps the Schmidt number,

$$\alpha_i^F := P_i / R\mathrm{Temp} \times V_i m_i$$

R is the gas constant in (J/m), Temp the temperature (deg K), P_i is the partial pressure (N/m^2), V_i is the molar volume (m^3/mol) and m_i is the molar mass (Kg/mol). This is a simplified version of the model that appears in [39].

Emulsification also affects evaporation and models for that interdependence have been proposed [36].

It is noted that these models have adequate predictive power for the first eight hours, but tend to over-estimate the rate at which oil evaporates beyond that time, particularly if the crude is very light (also, see [40]). Furthermore, [41] suggest that evaporation rates are greatly affected by wave action. Oil slick evaporation can thus exhibit aging. A comprehensive review and evaluation of the many models for oil spill evaporation is given in [42].

3.4.5. Photolysis, Biodegradation, Sedimentation

Photolysis initiates the polymerization and decomposition of complex molecules within a day after a spill occurs leads to chemical transformations that increase the solubility of the oil. This increases the oil's viscosity and promotes the formation of solid oil aggregates. In the oil slick Equation (7), the photolysis model is $-k_p^i s_i$, with $k_p^i \geq 0$. This simple rate equation, among other things, does not account for the reduction of UV rays due to cloud cover, daily variations in the ozone layer, effects of light scattering (see [43]).

Biodegradation affects both the surface as well as the bulk oil. Modeling biodegradation and sedimentation in the context of oil-fate dynamics is also crude at this stage. Biodegradation can be affected by both passively moving or actively moving biota. The biota itself has its own dynamic, which includes mortality and reproduction, and the population itself is affected by the oil by-products themselves. Accounting for losses due to biodegradation and sedimentation in our model will be done via a simple empirical loss rate. However, if the biota has a significant effect on the fate of oil it will also have to be explicitly modeled as a time dependent ecological system.

Sedimentation, will be modeled here as a simple first order mass loss rule, appears as a loss term in Equation (12). It can result from three processes: increased density of the oil as weathering proceeds, incorporation into fecal pellets via zooplankton ingestion or adhesion to or flocculation of the oil with suspended particulate matter. Sinking of oil through weathering alone is not expected in colder northern waters, although this has been observed in Gulf of Mexico and Persian Gulf blowouts. Numerous studies have been performed on the adhesion of oil to suspended particulate matter, but it remains difficult to adequately express the detailed dynamics of the process in a quantitative manner (see [44] and references contained therein). Sedimentation values can be as high as 30%.

Photolysis, biodegradation, and evaporation affect microscale properties of oil as well as the total mass of oil at the large scale of interest. An important future research question entails determining via a sensitivity analysis if the microscale effects could be folded into the empirical models for the evolution of the density, viscosity, and surface tension, so that only these phenomena can enter at the larger resolvable scales of the model.

4. Ocean Dynamics

At large spatio-temporal scales we account for wave effects via the Stokes drift. Hence, in what follows, we approximate the sea elevation by ζ^c. $\zeta^c = \hat{\zeta} + \zeta$ denotes the composite sea elevation. The sea elevation has been split into its dynamic component $\zeta(\mathbf{x}, T)$, and $\hat{\zeta}$, the quasi-steady sea elevation adjustment. $\hat{\zeta} = -A^2 k / (2 \sinh(2kH))$, where A is the wave amplitude and k is the magnitude of the wavenumber $\mathbf{k}$. The wave frequency σ is given by the dispersion relationship

$$\sigma^2 = gk \tanh(kH) \tag{20}$$

where g is gravity, and the evolution of the wave number is found by the conservation equation

$$\frac{\partial \mathbf{k}}{\partial T} + \nabla (\mathbf{k} \cdot \mathbf{v}^c + \sigma) = 0 \tag{21}$$

where $\mathbf{v}^c(\mathbf{x}, t) := (u^c, v^c)$ is the depth-averaged velocity (current) vector. The transverse directions are $\hat{\mathbf{x}}$ and $\hat{\mathbf{y}}$ for the across-shore and alongshore directions, respectively. The wave amplitude A is found by solving for the wave action

$$\mathcal{W} := \frac{1}{2\sigma}\rho_w g A^2 \qquad (22)$$

via the action equation,

$$\frac{\partial \mathcal{W}}{\partial T} + \nabla \cdot (\mathcal{W}\mathbf{c}_G) = -\frac{\epsilon}{\sigma} - \chi_S(\mathbf{x}, T)\nabla \cdot (D_w \nabla \mathcal{W}) \qquad (23)$$

The traditional wave action dissipation rate is captured by $\frac{\epsilon}{\sigma}$. The second term on the right hand side is another loss term which is the result of the presence of oil. $D_w \geq 0$ is the wave oil diffusion coefficient, which is different from Equation (8), and $\chi_S(\mathbf{x}, T)$ is the indicator function for the surface oil. The group velocity is $\mathbf{c}_G = \mathbf{v}^c + \mathbf{C}_G$, with $\mathbf{C}_G$ given by

$$\mathbf{C}_G = \frac{\sigma}{2k^2}\left(1 + \frac{2kH}{\sinh(2kH)}\right)\mathbf{k} \qquad (24)$$

The continuity equation is given by

$$\frac{\partial H}{\partial T} + \nabla \cdot [H\mathbf{V}] = 0 \qquad (25)$$

where $\mathbf{V} = \mathbf{v}^c + \mathbf{u}^{st}$. For monochromatic waves we can obtain the Stokes drift velocity via

$$\mathbf{u}^{st} := (u^{st}, v^{st}) = \frac{1}{\rho H}\mathcal{W}\mathbf{k} \qquad (26)$$

Following [45], in terms of the unidirectional spectrum $F(k)$ of a fully developed sea, the shallow water wave drift velocity is

$$\mathbf{u}^{st} \approx \frac{1}{2\pi g}(\cos\theta_0, \sin\theta_0)\int_0^\infty \sigma(k)^3 F(k)\frac{\cosh[2k(z+H)]}{\sinh[2kH]}\frac{d\sigma}{dk}dk \qquad (27)$$

for waves with local primary direction θ_0 and σ given by Equation (20).

The current velocity $\mathbf{v}^c$ is found via the momentum equation

$$\frac{\partial \mathbf{v}^c}{\partial T} + (\mathbf{v}^c \cdot \nabla)\mathbf{v}^c + g\nabla\zeta - \mathbf{J} = \mathbf{S} + \mathbf{N} + \mathbf{B} - \mathbf{D} \qquad (28)$$

The vortex force term (see [21]) is

$$\mathbf{J} = -\hat{\mathbf{z}} \times \mathbf{u}^{st}\omega \qquad (29)$$

where $\omega = v_x - u_y$ is the vorticity, and $\hat{\mathbf{z}}$ is the unit vector pointing anti-parallel to gravity. If coriolis forces are not ignorable, the term $\mathbf{J}$ is replaced by

$$\mathbf{J} = -\hat{\mathbf{z}} \times \mathbf{u}^{st}(\omega + 2\Omega)$$

All of the terms on the right hand side of (28) have alternative parametrizations as the ones that follow. The wind stress term

$$\mathbf{S} := \frac{1}{2H}C_D|\mathbf{v}^c|\mathbf{v}^c$$

with C_D the wind drag parameter. The bottom drag is

$$\mathbf{D} := \frac{C_M|\mathbf{v}^c|\mathbf{v}^c}{H^{5/4}} \qquad (30)$$

C_M is the Manning drag parameter, and the bulk dissipation is (see Section 3.3 and Equation (12)),

$$\mathbf{N} := \frac{1}{H}\nabla_\perp \left(\varphi D \nabla_\perp \mathbf{v}^c\right) \tag{31}$$

where φ is a proportionality constant. It is typical for the viscous dissipation and the oil dispersion to be qualitatively similar, and the momentum dissipation to be many times larger than the mollecular tracer diffusivity (high Schmidt number).

In the above expression we are assuming that the mechanical and tracer dissipation are proportional to each other. (This certainly is not an assumption in the model: the diffusivities can be far more independent from each other). Wave-to-current momentum exchanges due to the breaking waves are captured by

$$\mathbf{B} = \frac{\epsilon \mathbf{k}}{H \sigma} \tag{32}$$

There are several empirical descriptions of ϵ (≥ 0). The one we adopt here is due to [46]. (See also [47]). It is

$$\epsilon = 24\sqrt{\pi} g \frac{B_r^3}{\gamma^4 H^5} \frac{\sigma}{2\pi} A^7 \tag{33}$$

with B_r, γ, empirical parameters. This empirical relationship is based upon hydraulic theory and has been fit and tested against data in nearshore environments similar to the nearshore case considered in this paper.

5. Energy Conservation

Along with mass and momentum, the third conserved quantity of relevance to the transport model is the energy. It is included here for completeness, but its utility as an important constraint that generates equations of state is not discussed here. We denote the energy density, per unit (transverse) area as

$$\mathcal{E} = \frac{1}{2}\rho \mathbf{V}\cdot\mathbf{V} + \frac{1}{2}g\rho_w \eta^2 + \rho e$$

where e is the internal energy. The first three terms are associated with mechanical work, the last one is the internal energy associated with the 2 oil compartments. ρ_w and ρ are the ocean water and the oil mass densities. The conservation of energy equation is

$$\left[\frac{\partial}{\partial T} + \mathbf{V}\cdot\nabla\right]\mathcal{E} = -Y + \frac{1}{H}\int_{-h}^{\eta}\left(\mathbf{F}_b\cdot\mathbf{U} + \nabla\cdot(\sigma\mathbf{U}) + \rho\mathbf{Q} - \nabla\cdot\mathbf{K}\right)dz$$

The first term on the right hand side accounts for the kinetic energy in the unresolved scales, *i.e.,* $Y = \frac{1}{2}\rho\mathbf{V}'\cdot\mathbf{V}'$, where $\mathbf{V}'$ is the velocity fluctuations associated with the depth-averaged Reynolds stresses and the depth-averaged difference between the depth- and wave- averaged velocities. The next two terms correspond to body and surface (mechanical) forces. The last two terms include the thermal, electromagnetic, and chemical sources/sinks and fluxes.

6. Illustrative Dynamic Examples

In this paper we limit ourselves to highlighting phenomena that owe their peculiarities to the advection and the dispersion of the oil model. In [26] we illustrate details of the mass exchanges and thus will use a particularly simple mass conserving mass exchange term in the calculations, and in [48] we present the dimension reduction properties of the aging effects of the chemical complexes.

In the first computational example we revisit the issue of nearshore sticky waters, described in [49]. Nearshore sticky waters refers to the apparent slowing down and possibly parking of buoyant pollutants, beyond the break zone, as they travel toward the shore from deeper waters. In the second example we highlight dispersive effects associated with the parametrization that includes effects due to the waves.

6.1. Nearshore Sticky Waters in Shores with Intense Breaking

In [49] we proposed an explanation for the apparent slowing down and parking of inconing buoyant tracers, in the neighborhood of the surf zone. We labeled this slowing down as the nearshore sticky water phenomenon. One of the outcomes of this work is that eddy and turbulent dissipation near the shore leads to a thickening of the mixed layer and a stalling of the average flux due to advection due to the currents, the flux becomes diffusive in nature. In that work we only considered a conceptual model, and focused only on dispersion in the tracer. In this study we revisit this problem to consider wave momentum transfers to currents and dispersion in both the tracer and the ocean flow. The specific aim of the following calculations is to compare the effect of different advection and dispersion models on nearshore sticky waters.

Figure 5 depicts the physical domain.

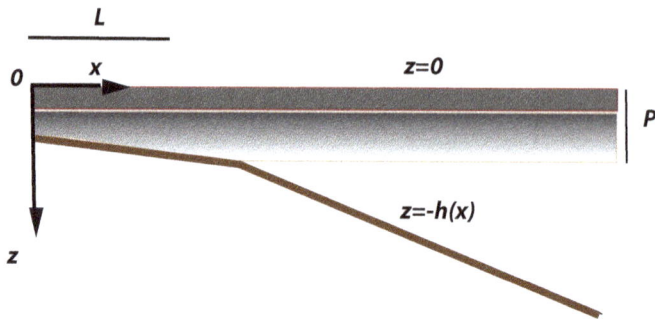

Figure 5. Schematic cross-section of the model domain. A light, thin oil slick sits atop the ocean. The ocean's mixed layer of thickness P is laden with oil droplets, accounted for as a concentration. The distance from the shore, at $x = 0$, is denoted by x. The break zone extends to $x = L$. The ocean surface is at $z = 0$ and bottom topography is fixed and described by $z = -h(x)$.

The quiescent ocean level is at $z = 0$, the basin is bounded below, at $z = -h(x)$. The domain extends from $x = 0$, the shore end, where the depth is $h_0 \geq 0$, to $x = x_m$ where the depth is $h_m \geq x_0$. The bathymetry $h(x)$ will be sloped and featureless:

$$h(x) = h_0 + mx, \quad 0 \leq x \leq x_m$$

where h_0 is the depth in the nearshore, and $m \geq 0$ is the slope. We distinguish two oceanic regimes in our problem: the high mixing surf zone, corresponding to $0 \leq x \leq L$, and the deep ocean zone, from $L < x \leq x_m$. L is typically tens to hundreds of meters. The pollutant (for example, oil), or the tracer (for example, an algal bloom) is subject to buoyancy effects. Oil in the surface slick may be entrained by the action of wave breaking and turbulent mixing. The oil may also resurface, at a rate dependent on the size of the droplets. We will assume that the *oil slick* has thickness $s(x, t)$ per unit length, typically micrometric. Immediately below is a layer of ocean in which the bulk of the oil is found, in suspension. As depicted in Figure 5, the layer containing the suspended oil is assumed to have a maximum thickness P, as given by Equation (14). The subsurface oil has an effective thickness $S(x, t)$ per unit length. Assuming that the interior oil is uniformly distributed within the mixed layer, we have the *equation of state*

$$S(x, t) = C(x, t)\xi(x) \tag{34}$$

where C denotes the (dimensionless) volume fraction of the oil in suspension (see Equation (10)), and $\xi(x)$ is the local depth of the mixed layer. We approximate $\xi(x)$ as a smooth approximation to $\min(h(x), P)$. The waves are coming in at an angle θ with respect to the normal to the shore. Typically θ is less than 10 degress. A geophysically-inspired value for θ is 3^0 (*cf.* National Data Bouy Center, for the Middle Atlantic Bight, e.g., Duck NC USA). Typical as well is that the waves are about 1m high about 1Km away from the shore. These waves generate a alongshore current (see [50] and references contained therein). The ocean flows might include residual flows due to waves as well as wind-induced and gradient flow-induced currents. However, we focus on the most fundamental of situations, namely, flows due to waves. We assume the incoming waves generate a Stokes drift $\mathbf{u}^{st} = (u^{st}, v^{st})$.

We assume steady conditions, ignore alongshore variation and take $H \sim h$. Also, the group and wave phase speed are $c_G \approx \sqrt{gh}$. Since $u^{st} + u = 0$, the cross-shore momentum, as given by Equation (28), is

$$\hat{x} \cdot (-g\nabla\zeta + \mathbf{B} + \mathbf{N}) \approx 0$$

which describes the wave setup. In the alongshore direction we obtain the balance $\hat{y} \cdot (-g\nabla\zeta + \mathbf{B} + \mathbf{N}) \approx 0$, from which one can obtain an equation for the alongshore velocity. Assuming a linear drag model,

$$C_M v - \alpha_b k \sin\theta \frac{A^7}{h^5} + \hat{y} \cdot \mathbf{N} = 0 \tag{35}$$

with C_M the bottom drag parameter. The second term in Equation (35) is $\hat{y} \cdot \mathbf{B}$ and represents the transfer of momentum to currents by the residual stresses due to breaking in the nearshore (see [46]). The parameter $\alpha_b = 12/\sqrt{\pi g}B_r^3/\gamma^4$, where we set $B_r = 0.8$ and $\gamma = 0.43$ in the calculations that follow. The lateral eddy viscosity is

$$\hat{y} \cdot \mathbf{N} = \frac{\partial}{\partial x}\left(K\frac{\partial v}{\partial x}\right)$$

where

$$K = \begin{cases} 0.02\sqrt{gh(x)}h, & 0 \leq x \leq L, \\ 0.02\sqrt{gh(L)}h(L)\left(0.8[\frac{h(x)}{h(L)}]^{-4} + 0.2\right), & L \leq x \end{cases} \tag{36}$$

This eddy viscosity model is a version of a model (see [51]) that is commonly used in the nearshore engineering community (see [52], and references therein).

As shown in [50] and [53], if the waves are assumed steady, an approximate solution to Equation (23) is possible. (The loss rate due to the presence of oil is ignored here). The wave amplitude is then given approximately as

$$A(x) = h^{-1/4}[h_m^{-5/4}A_m^{-5} - \tilde{\delta}(h^{-23/4} - h_m^{-23/4})]^{-1/5} \tag{37}$$

with $\tilde{\delta} = \frac{10\delta}{23s}$. Here, $\delta = 2\alpha_b\sigma g^{-3/2}$. The depth $h_m = 8$ m and $A_m = 0.8$ m. The Stokes drift velocity is given by

$$u^{st} = -\frac{A^2\sigma k}{2\sinh^2(kh)}\cosh[2k(z+h)] + \frac{A^2\sigma \sinh(2kh)}{4h\sinh^2(kh)} \tag{38}$$

(The minus sign is due to its shoreward direction). The second term represents the undertow (see [54] for a discussion on the undertow in the nearshore). This Stokes drift velocity is consistent with the kinematic constraint that the depth-mean shoreward velocity must be equal to zero.

We consider the simplest possible situation: no wind, no sources/sinks of oil, we do not invoke reactions or biodegradation. Further, we set $C_{xs} = 1$. Equations (7) and (12) become

$$\frac{\partial s}{\partial T} + \frac{\partial [u^{st}(x)s]}{\partial x} = -\frac{(1-\gamma)s - \gamma PS}{\varsigma(x)} + \frac{\partial}{\partial x}\left[\Psi \frac{\partial s}{\partial x}\right] \tag{39}$$

$$\frac{\partial S}{\partial T} + \frac{\partial [u_e(x)S]}{\partial x} = \frac{(1-\gamma)s - \gamma PS}{\varsigma(x)} + \frac{\partial}{\partial x}\left[\Psi \frac{\partial S}{\partial x}\right] \tag{40}$$

where we have used Equation (34) and

$$u_e(x) := u_C(x) + \Psi \frac{1}{\xi(x)} \frac{d\xi(x)}{dx} \tag{41}$$

is the *effective* subsurface oil velocity and $u_C(x)$ is the ξ-averaged velocity of the subsurface oil. We also use the same simple mass exchange model for R used in [49]. This is the first term on the right hand side of Equations (39) and (40). In R the constant γ captures the propensity of oil to exchange between the slick and the subsurface, and ς is a measure of the rate of the exchange, which is proportional to $1/k^2\Psi$. In [49] we used a crude model for the slick and the subsurface oil velocities. The Stokes drift velocity was approximated by a parabolic profile in z with constant values in x. In the calculations that follow we instead use Equations (38) and (41) for the slick and the subsurface velocities, respectively. In this study we use

$$u_C = \frac{1}{\xi} \int_{-\xi}^{0} u^{st} dz = -\frac{A^2\sigma}{2\xi h(\mathcal{Z}_h^2 - 1)^2}\left[\mathcal{Z}_h^4(h - \xi) + h\mathcal{Z}_\xi^2 - h\mathcal{Z}_h^4\mathcal{Z}_\xi^{-2} - h + \xi\right]$$

where $\mathcal{Z}_* := \exp(k*)$.

The boundary conditions at the shore and the far end of the domain are:

$$\begin{aligned} u^{St}(x)s - \Psi\frac{\partial s}{\partial x} &= 0 \text{ at } x = 0 \text{ and } L, \\ u_C(x)S - \Psi\frac{\partial S}{\partial x}S &= 0 \text{ at } x = 0 \text{ and } L \end{aligned} \tag{42}$$

With initial conditions, these equations become a well-posed problem.

In [49] the ad-hoc model for the dispersion $\Psi = \Sigma + \Xi \approx D$ was

$$D = D_{eddy} + \mathcal{S}(x)D_L \tag{43}$$

the wave dispersion has $D_L = 1.6 \text{ m}^2/\text{s}$, $\mathcal{S}(x) = (1 + \exp[(x - L)/w])^{-1}$, where $w = 20$ m is the transition width, $L = 200$m. The turbulent eddy viscosity is constant: $D_{eddy} = 0.05 \text{ m}^2/\text{s}$.

In the first computational example we will compare the outcomes of using the dispersion used in [49], Equation (43), with those using

$$\Psi = 1.6K \tag{44}$$

where K is given by Equation (36). The dispersion Equation (44) is familiar to the nearshore community and it is this reason why we want to make a comparison of the results to those obtained using the ad-hoc dispersion, Equation (43).

In the following calculations, $X = 1000$m, $L = 200$m, $h_0 = 2$m, $h_\infty = 100$m, $\gamma = 0.1$. Figure 6a shows the initial condition for $s(x,0)$. The initial condition on S is zero. $P = 3$, which means that the intersection between the bottom topography and P is at around 165 m from the shore. Figure 6b shows the ampltidue of the wave, with $A_m = 0.8$m, and $h_m = h_\infty$, $\sigma = 2\pi/9.1$ radian/s, (see Equation (37)). Figure 6c compares the $u_e(x)$, see Equation (41) as obtained when using Equations (43) and (44). Both u_e have similar zero crossings, at roughly 240 m, and are similar in the deeper reaches of the domain. However, they are different close to the shore: clearly, it is $\Psi(x)\frac{1}{\xi(x)}\frac{d\xi(x)}{dx}$ that yields the behavior of the effective subsurface velocity in the shallow end of the domain. Figure 6d compares the dispersions.

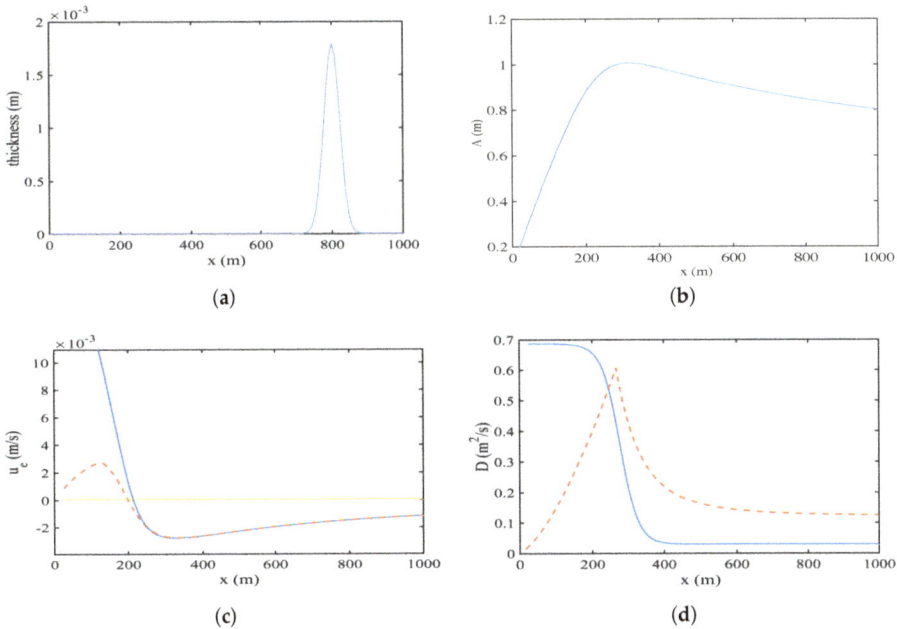

Figure 6. (**a**) The initial pulse $s(x,0)$. $S(x,0) = 0$; (**b**) Wave amplitude $A(x)$; (**c**) Effective velocity $u_e(x)$, using (Equation (43) (solid) and Equation (36) (dashed)); (**d**) Comparison of the diffusion models, Equation (43) (solid) and Equation (36) (dashed).

Figure 7 shows the space-time evolution of $S(x,t)$ with the two different choices of dispersion, Ψ. Figure 7a,b correspond to the outcomes using Equation (43), whereas Figure 7c,d correspond to using Equation (44). Both cases are qualitatively similar and demonstrate that even under more realistic modeling assumptions, the results from [49] still hold. However, going beyond the cases considered in the prior paper, the Stokes drift velocity intensity gets larger as the waves approach the break zone, then the waves transfer momentum to the generation of the longshore current due to breaking. Hence, the enhanced mixing is not only affecting the tracer, it is also affecting the mechanics of the currents and waves in the case portrayed here.

Specifying P constant is very unrealistic. A second illustration allows us to consider what happens when the mixing layer depth $P = P(x)$. In this case we consider $h_0 = 2$ m, $h_\infty = 100$ m. $P = 1/k(x)$. All other parameters remain the same, and we use the Equation (44) dispersion. Figure 8a shows the bottom topography, and superimposed, $P(x)$. The two curves cross at around $x = 195$ m, which is an estimate of L. We note that P and L are not independently specified in this case, unlike the previous one. The wave amplitude appears in Figure 8b. Figure 8c displays u^{st}, u_C. Figure 8d displays the dispersion.

Figure 9 shows the effective subsurface oil velocity $u_e(x)$. The zero crossing is at roughly 480 m away from the shore.

(a)

(b)

(c)

(d)

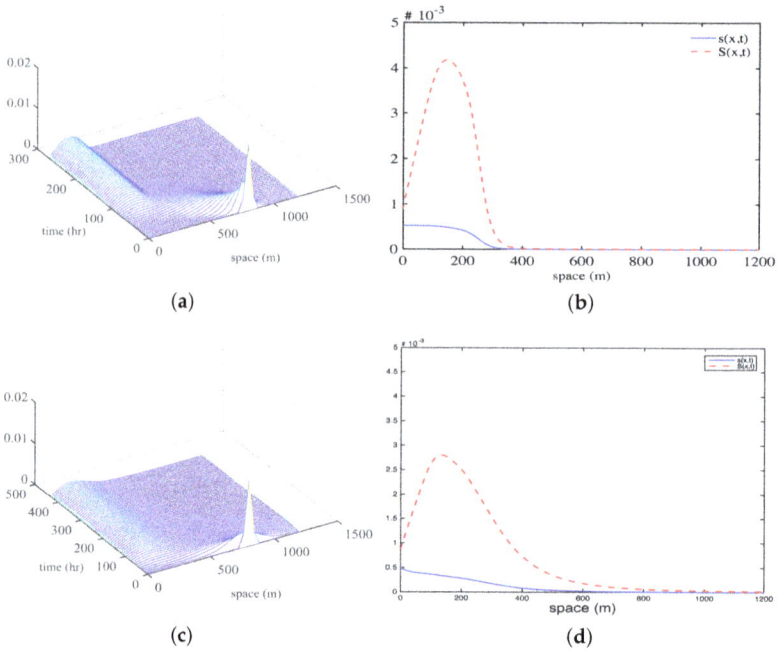

Figure 7. (a)–(c): $S(x,t)$, (b)–(d): the cross section of S at the final time; (a), (b) used Equation (43) dispersion in the calculation, (c), (d) computed using Equation (36) dispersion.

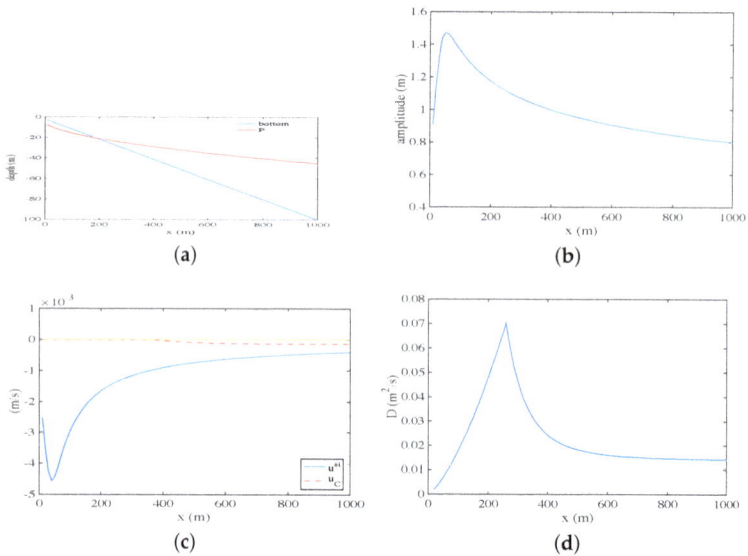

(a)

(b)

(c)

(d)

Figure 8. (a) Bottom topography $h(x)$ and $P(x)$; (b) wave amplitude $A(x)$; (c) oil slick velocity u^{st} and subsurface oil velocity u_C; (d) dispersion $0.16K(x)$.

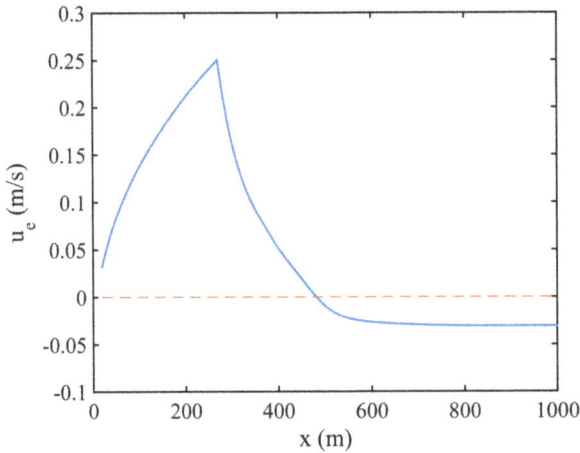

Figure 9. Effective subsurface velocity $u_e(x)$. The advection associated with the dispersion dominates. The crossover from positive to negative occurs approximately at $x = 480$ m.

In Figure 10 we display the outcomes for $S(x,t)$.

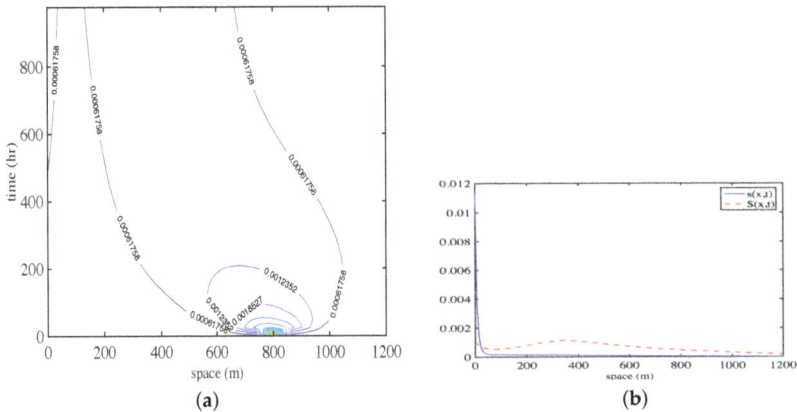

(a)

(b)

Figure 10. (a) Evolution of contours of $S(x,t)$; (b) at $t = 1000$ h, the distribution of s (solid) and S (dashed).

The bulk of the oil, which started in the slick, quickly moves to the subsurface. Whatever oil is in the slick makes it to the shore. However, the subsurface oil stalls, its center of mass is slightly over 400 m away from the shore, where u_e crosses zero (see Figure 9). Speculation was that the stickyness results in [49] were critically dependent on the magnitude and the shape of the dispersion. In this study we show that even with more acceptable dispersion parametrization models, which also happen to be less dispersive near the shore, parking and slowing down of the advection of oil toward the shore is possible. Furthermore, we also make use of more realistic conditions on the hydrodynamics. To conclude we mention further that we focused on across-shore dynamics, but it should not be forgotten that the oil also will travel alongshore due to the longshore current v, which appears in Equation (35). This added dynamic does not modify our conclusions regarding nearshore stickyness.

6.2. Shelf Dynamics Examples

We continue the exploration of advective and diffusive outcomes of the model. The examples highlight the critical role played by the Stokes drift velocity in oil transport. As we discussed previously in Section 3.2, the waves participate in the advection of tracers, via the Stokes drift velocity, and in the diffusion term. The goal of the following calculations is to suggest three ways in which the wave components play important roles in the transport of oil. For this purpose we will forgo the interaction of the slick and the subsurface oil components, and examine how a tracer $C(\mathbf{x}, t)$ evolves according to Equation (12).

The Stokes drift velocity plays an important role in the dynamics at, and below, the sub-mesoscale (See [55], for details and references). At these scales fronts, filaments and baroclinicity are evident, as is Langmuir turbulence. In the following examples we take inspiration from ocean conditions near the shores of the Gulf Coast. Figure 11 shows the region. We focus on an intermediate scale: days and tens of kilometers. At these intermediate scales we will not see familiar mixing and smoothing of small scale features by diffusion processes, nor will we see the processes that are more dramatic at the large scales, such as complexities induced in the oil distribution due to large scale flow features associated with basin and bottom topography and winds, and barotropic effects. In this region the bathymetry is not changing greatly and the wind is fairly uniform across the extent of the domain; however, the wind does changes significantly in time. According to [56], the season-averaged winter and early summer steady currents in the region under consideration are approximately 0.15 m/s and 0.05 m/s and directed at $-100°$ and $45°$ with respect to true North, respectively.

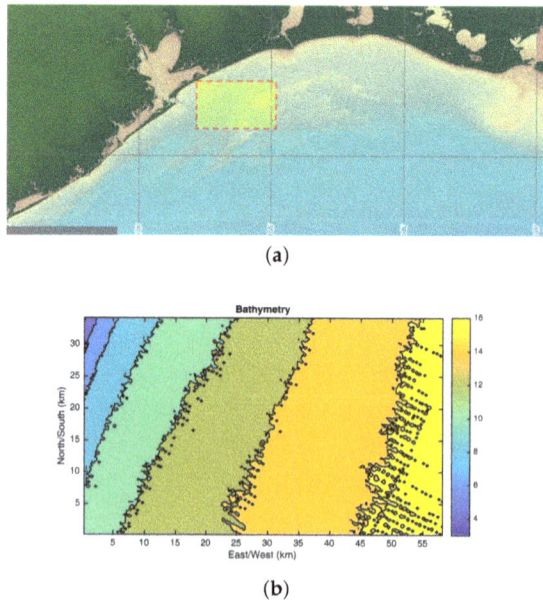

(a)

(b)

Figure 11. (a) Areal view, with the domain of the computation, highlighted. South Etast of Galveston, TX, USA; (b) bathymetry of the region [57].

Wind data is used to estimate the Stokes drift, via Equation (27). As it turns out, the winter currents are larger in magnitude than the Stokes drift, however, they are comparable in the summer months. For wind data we avail ourselves of the January–December 2010 wind data from www.ndbc.noaa.gov/, Station 42035, off of Galveston TX. The wind data is available in 10 minute intervals. In what follows we use the steady currents described above, calling these the (seasonal) winter and summer currents and a Stokes drift velocity computed from wind data. The Stokes drift velocity will be time dependent and only mildly position dependent. With regard to the dispersion, we apply the parametrization Equation (15) for Ψ with $\Sigma \approx 0$. The part of Ξ associated with currents is constant, however, mildly spatially dependent since it depends on H. The typical sizes of the components of Ξ in Equation (17) is 0.01 m²/s for $\kappa\delta_{ij} + \beta\Theta H$, and 0.16 m²/s for the dispersion associated with waves, $\beta\Theta^{st}/k$.

Changes in the Stokes drift, due to changes in wind direction, leads to a time dependency of the advection and of the diffusion terms.

Figure 12 shows the path of an ideal tracer subjected to the Stokes drift. The drift is updated every 10 min and the tracer path represents approximately 200 h of wind. The path generated by the wind-induced Stokes drift is complex. If we traced several paths, with different starting points, we would see qualitative similarities among them, but nothing dramatic. The changes are due to the depedency of the drift on the water column depth which introduces slight local differences to the drift velocity at any given time, throughout the domain.

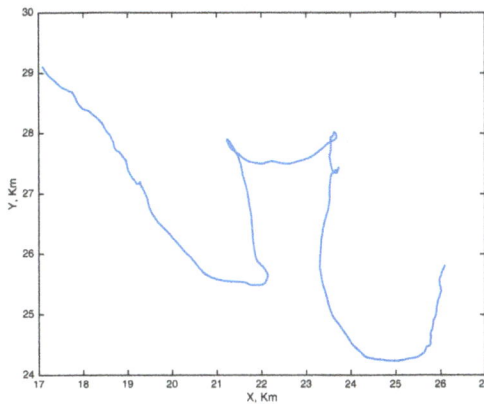

Figure 12. Lagrangian path of an ideal tracer induced by the Stokes drift from the wind data. The path would be identical in all cases in this section in which we are activate the wave-induced advection. The path direction in time is to the right.

In the first two examples we will discuss, we use the Galveston shore bathymetry and the abovementioned wind data to generate the Stokes drift. We, however, replace the currents with a synthetic shear current $\mathbf{v}^c$ with maximum magnitude of 0.5 m/s. The shear current is constant in time and is displayed in Figure 13a. The initial distribution $C(\mathbf{x}, 0)$ is chosen to be a combination of two Gaussian functions, one of which has been multiplied by a random uniform amplitude (see Figure 14a). The conditions at the boundary for the tracer are periodic (the flow and the bathymetry is periodized). In Figure 14b we display $C(\mathbf{x}, t)$ after about a week. The conditions for this run are: we applied a shear $\mathbf{V} = \mathbf{v}^c$ and no Stokes drift. The diffusion has no wave-induced contribution (*i.e.*, $\Xi = 0$). Figure 14c can be compared with Figure 14b to get a sense of how much the added wave-induced contribution (*i.e.*, $\Xi \neq 0$) afftects the tracer smoothness. Figure 14d shows the more complete dynamics of wave-induced advection and shearing, $\mathbf{V} = \mathbf{v}^c + \mathbf{u}^{st}$, and a diffusion tensor that includes wave effects; both terms in Ξ are non-zero. The overall general observation is that wave-induced effects are significant in determining the fate of the tracer with regard to position and structure, in the latter respect we see smoothing at scales in the order of a kilometer or less, as would be expected from simple dimensional estimation.

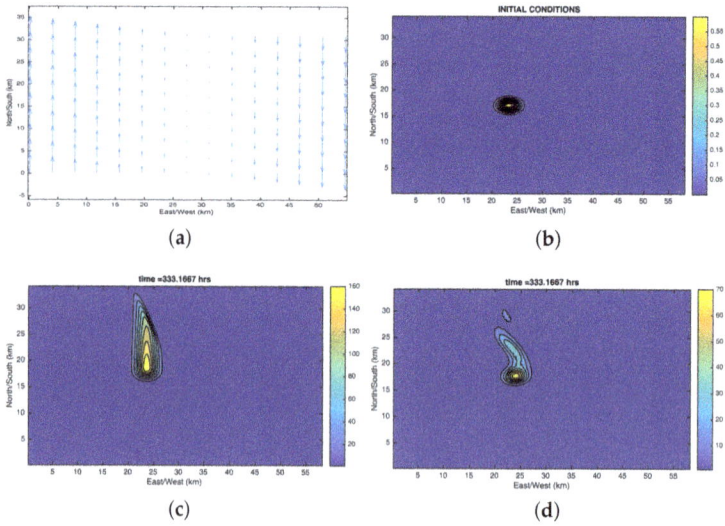

Figure 13. Waves can localize a spill: (**a**) the steady currents; (**b**) The source location. (**c**) Spill under the action of the shear flow; (**d**) spill under the action of shear and Stokes drift advection.

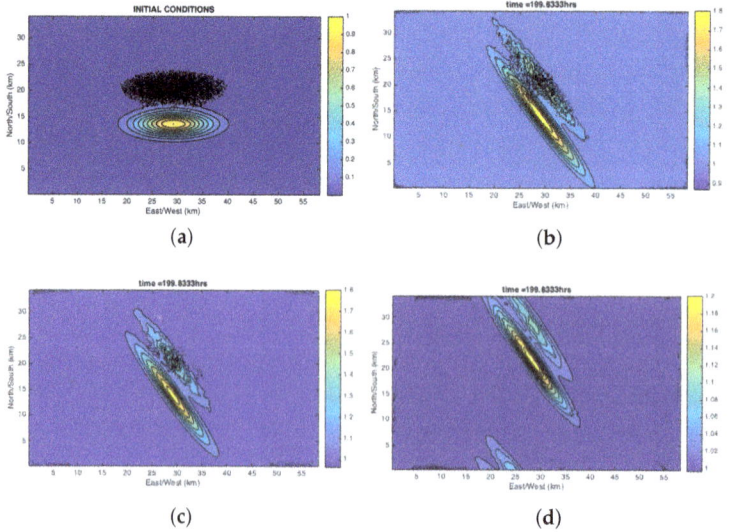

Figure 14. Effect of wave-driven diffusion/dispersion: (**a**) Initial conditions. Two Gaussians, the upper one has been multiplied by a uniformly random amplitude. Final time configurations: (**b**) Constant shear velocity applied as shown in Figure 13a, Ψ does not include waves and the advection is due to the shear only. (**c**) Shear advection and diffusion includes wave component; (**d**) Shear and waves included in the diffusion and the advection.

In the next example we place a steady source of oil at locations $(x, y) = (0.4L_x, 0.5L_y)$. The source will produce Figure 13a initially (the initial conditions for C, however, are zero). The advection $\mathbf{V} = \mathbf{v}^c + \mathbf{u}^{st}$, where the current is the shear shown in Figure 13a, and the Stokes drift velocity corresponds to the first 333 h of wind of the 2010 year. The diffusion is the full tensor Ψ. We first show in Figure 13c the tracer's fate, after 333 h, with the Stokes drift advection suppressed. With the Stokes drift velocity added, the outcome at the same time is shown in Figure 13d. The Stokes drift localized the tracer to the neighborhood of the source location.

Figure 15 shows the difference in the empirical dispersion, as estimated by the mean square distance, which we denote $(\text{msd}_x(t), \text{msd}_y)(t))$. With $\mathbf{r} = \mathbf{x} - \mathbf{x}_0$, where $\mathbf{x}_0$ is the location of the source,

$$(\text{msd}_x, \text{msd}_y) = \frac{\int_\Omega (\mathbf{r} - \langle \mathbf{r} \rangle)^2 C(\mathbf{x}, t) dx dy}{\int_\Omega C(\mathbf{x}, t) dx dy} \tag{45}$$

where $\langle \mathbf{r} \rangle$ is the time-dependent centroid

$$\langle \mathbf{r} \rangle(t) = \frac{\int_\Omega \mathbf{r} C(\mathbf{x}, t) dx dy}{\int_\Omega C(\mathbf{x}, t) dx dy}$$

Usign the mean square distance once can compute an empirical time dependent variance as the L_2 norm of the mean square distance. The mean square distance associated with the case depicted in Figure 13 is shown in Figure 15.

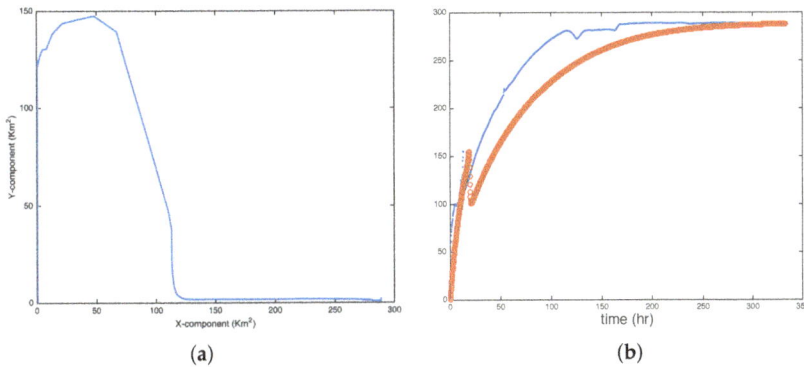

(a) (b)

Figure 15. Mean square distance ($\text{msd}_x, \text{msd}_y$) evolution, given by Equation (45), associated with case shown in Figure 13d; (b) empirical variance, as a function of time, associated with Figure 13c (thick line) and Figure 13d (thin line).

Figure 15 suggests that the variance of the tracer increases very fast in the vertical direction in the beggining. The tracer variance at longer times increases mostly in the horizontal direction. The pause that is evident in the empirical variance, shown in Figure 15, is associated with the decrease of the y−component of the mean square distance. The history of the variance with and without waves is different with a tendency of the case with no waves to reach the value similar to the case with waves much slower.

In the next two illustrations we will use the field-inspired seasonal winter and summer steady currents, the data-driven Stokes drift. However, we will make the domain larger. The overall depth of the basin is set to 14 m. We use a steady tracer source located in the center of the domain. In Figure 16 we contrast the difference in tracer evolution with and without the wind-induced drift velocity. Figure 16a is the case with winter currents only, and Figure 16b has the added advection due to the Stokes drift velocty. Shown is the tracer after nearly 200 h. The tracer source is located in the center of the domain. The winter current dominates over the Stokes drift and thus the differences are subtle. It is noted, however, that the maximal tracer density is not at the source location.

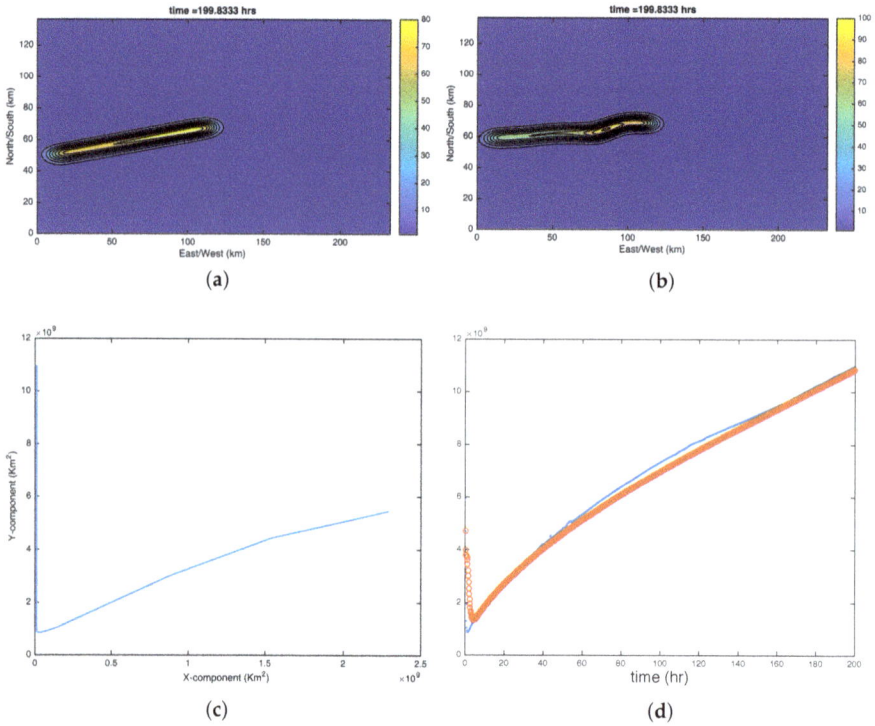

Figure 16. Tracer at the final time due to a steady source, located in the middle of the domain. Winter conditions. (**a**) Currents only; (**b**) Currents and Stokes drift. (**c**) Mean square distance corresponding to case (**b**); (**d**) Empirical variance corresponding to currents only (thick line), and currents and waves (thin line). The currents are overwhelming and thus the waves have a minor effect on the evolution of the spill, mostly at short times. Advection due to Stokes Drift velocity only. The time interval between points is 10 min.

As a last illustration, we consider the same domain but now we invoke the (weaker) summer current which is directed toward the north-east. Figure 17a shows the final tracer distribution due to a point source under the action of the steady summer current, after over 333 h. The dynamics of the same tracer with the Stokes drift velocity is dramatically different: see Figure 17b. Not only is the plume structurally different when the current and the Stokes drift are invoked, the location of highest tracer concentration are not at the source location. Hence, the effect of both advection and diffusion due to the waves is critical. With higher resolution dynamics the meandering plume is more jagged at fine scales. Nevertheless, the meandering aspect is very much a typical distribution of small to mid-size oceanic oil spills, as is the fact that the largest concentration of oil is not necessarily concentrated at the source location.

(a) (b)

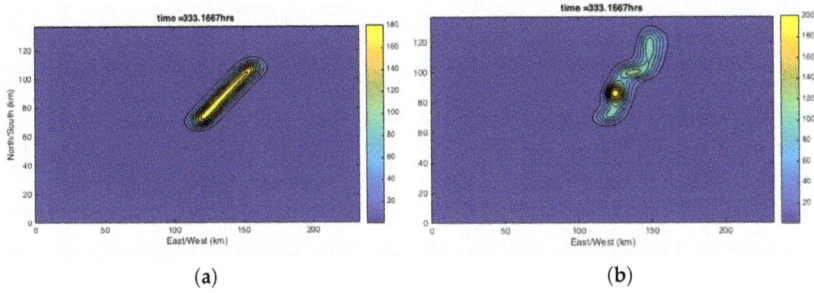

Figure 17. Summer conditions. Final configuration of tracer due to a point source located in the center of the domain, after about 333 h. (**a**) Steady current, no Stokes drift velocity; (**b**) steady currents and Stokes drift.

Figure 18 shows the evolution of the summer plume, at regular time intervals.

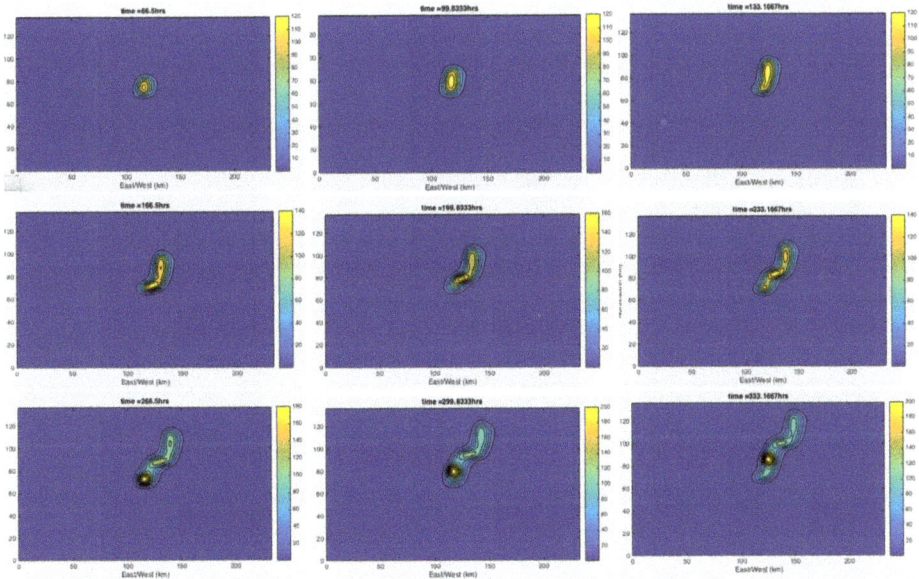

Figure 18. Summer conditions, currents and waves. Evolution of tracer leading to Figure 17b. The current is steady at 0.18 m/s, directed in the North East direction. The wave-induced Stokes drift and the currents are comparable in magnitude.

7. Recapitulation

The long term goal of this project is produce a model and simulation tool that captures, with reasonable accuracy, the distribution and evolution of oil (or a similar pollutant) due to an oil spill, over large spatio-temporal scales, typical of those required by hazard and policy studies. Modeling, computational, and engineering novelties generated by this project should apply to the modeling of similar pollution problems. Our modeling strategy favors techniques and choices that lead to a mass-conserving model. A faithful model for the oil dynamics, along with all of the complexities of its interconnections to oceanic, bio-geochemical, and atmospheric dynamics, is a long-term undertaking.

In this work we highlighted aspects related to the advection and the dispersion of oil. The oil dynamics model consists of a surface oil slick and a subsurface oil component. The spatial slick footprint is defined by a thresholded concentration of oil on the sea surface *i.e.*, surface oil is near or at the surface and of sufficiently high concentration to be approximated by pure oil. The subsurface oil is modeled instead by a concentration.

Oil has many chemical components and these react with each other. Numerically capturing the transport of oil means having to contend with high dimensions and with reaction rates that lead to very stiff time integration challenges. The dimension reduction will be achieved by defining low-dimensional chemical complexes. However, the introduction of these complexes produces unusual chemistry among the complexes, with non-autonomous chemical reactions. We borrow the term *aging* or *weathering* to connote this effect and its impact extends beyond chemistry to such processes as sedimentation, evaporation, photolysis, and biodegradation. We will be using a stochastic parametrization and projection methods to capture the oil in terms of a handful of oil chemical complexes. Weathering and its parametrization is presented in a companion paper [48].

At the very large spatio-temporal scales we are focusing on physical processes that are germane to microdynamics cannot be feasibly resolved. However, these processes are unavoidable. The interaction of the surface and subsurface oil necessitates the incorporation of microscale physics in the model. For example, capturing the time evolution of viscosity and surface tension is essential. At present the plan is to use empirical models for the time evolution of viscosity and surface tension. Both of these material properties affect the evolution of the droplet distribution equations that make up the subsurface concentration. We keep track of the droplet distribution because it is essential to getting the mass exchange interactions between the slick and the subsurface oil, and hence, to mass conservation. The details on the mass exchange dynamics, along with the upscaling strategies that circumvent resolving microscales, albeit with a loss of fidelity appears in [26].

In the future phenomenology that is crucial to using the model for hazard studies will be adopted. Particularly important are the biodegradation and the photodegradation/evaporation/sedimentation. Source characterization is also critical, particularly when the source of the oil is located on the sea bottom. The plan is to adapt existing characterizations of these due to others.

In this study we highlighted the more fundamental processes of advection and diffusion/dispersion. In particular, the study shows how wind, waves, and currents affect the advection of oil. In one of the illustrations we consider the slowing or parking of oil traveling toward a beach. We call this phenomenon *nearshore sticky waters*. We proposed the basics of the mechanism in [49] using a conceptual approximation of the oil model presented herein. We took the opportunity in this paper to revisit this problem using more faithful models for the velocity, the waves, and the diffusivity. We also derive an expression for the depth of the mixed layer which in the original paper was an input parameter. The outcomes in this study confirm the conclusions reached using a conceptual model in [49] using more realistic models for elements critical to the explanation of the phenomenon.

Pictures of oil spills at wave scales indicate that waves and wind have high impact on the topology and evolution of oil slicks via Langmuir turbulence and vertical mixing (see [55]). We showed that at scales larger than those of waves, currents, and wind are all important in the mechanics of oil. Effects, such as increases in oil dispersion as well as dispersion suppression are possible, as are the generation of tortuous advection-dominated distributions of oil slicks and concentrations, much of this driven by the fact that the time scales of wind, the residual flow due to waves, and currents span a wide range. Furthermore, at very large scales transverse, eddy-scale dissipation produces a smoothing of smaller scale features in ways that are not surprising, however, diffusivity is due to wall and bottom driven turbulence as well as wave-driven mixing due to waves. The scale at which advective and diffusion/dispersive effects are bound to be most dramatic is at sub-mesoscale scales [58,59]. At this scale frontogenesis and filamentation as well as Langmuir and background turbulence play important roles. Moreover, at these much larger scales wind, waves, and currents are affected in very strong ways by topographic/bathymetric effects, as well as barotropic balances. The dynamics of oil as captured

J. Mar. Sci. Eng. **2015**, *3*, 1504–1543

by our model at the sub-mesoscale will have to wait till the physical model is implemented in a high performance ocean/atmosphere/waves circulation model. The plan, presently, is to integrate the transport into an already existing circulation model for the shelf waters of the Gulf of Mexico, such as ROMS (see [60]) with a wave component provided by Wavewatch. The atmosphere will be simulated via The Weather Research and Forecasting Model (WRF). The simulation platforms are already in use in coupled form.

Finally, in this study we were also explicit about the equations of the ocean flow dynamics, as well as the energy conservation equation, consistent with the depth-averaging and the spatio-temporal scales of the oil transport model. Of note the oil contributes a dissipation term to the wave action, in locations where oil is present. The effect is to suppress high frequency components of the wave spectrum. It might be the case that the presence of significant amount of oil at the sea surface affects the coupling of wind stresses to the ocean flow, however, we did not make this speculation explicit here. At seconds-Km scales, where LES methods can be used to capture Langmuir turbulence the presence of significant amounts of oil should have an effect on the instabilities that lead to Langmuir circulation, since oil affects the wave spectra and might also affect the stratification in the wave boundary layer.

Several oil transport models are based upon Lagrangian tracer dynamics. We opted for an Eulerian description because we see less practical challenges in achieving mass conservation this way. The Lagrangian model, however, does not have to distinguish between surface and subsurface oil, if full three dimensional flow simulations are used, but do require an interpretation if depth-averaging is invoked in the ocean dynamics. Our conclusions regarding advective and diffusive processes presented here should have a bearing on Lagrangian based models. It is the interplay and analysis of both approaches that will eventually lead to a practical and accurate simulation capability.

Acknowledgments: This work was supported by GoMRI and by NSF DMS grants 1524241, 1434198 and 1109856. JMR also acknowledges The NSF funded Institute of Mathematics and its Applications, as well as the J. T. Oden Fellowship at the University of Texas, Austin.

Author Contributions: Conceived and designed the model: Juan M. Restrepo, Jorge M. Ramírez and Shankar Venkataramani. Analyzed the model: Juan M. Restrepo, Jorge M. Ramírez and Shankar Venkataramani. Wrote the paper: Juan M. Restrepo, Jorge M. Ramírez and Shankar Venkataramani.

Conflicts of Interest: The authors declare no conflict of interest.

Abbreviations

Name	Symbol	Units
fast/slow time	t, T	s
transverse position vector. Cross-shore, along-shore coordinate	$\mathbf{x} = (\mathbf{x}, \mathbf{y})$	m
depth coordinate	z	m
cross-shore, along-shore unit vectors	$\hat{\mathbf{x}}, \hat{\mathbf{y}}$	-
sea elevation	$\eta = \zeta^w + \zeta^c + S$	m
bottom topography, total water column	$h, H = h + \eta + S$	m
spatial gradient operator	$\nabla = (\nabla_\perp, \partial_z)$	1/m
wave and mean (current) sea elevation	$\zeta^w, \zeta^c = \zeta + \hat{\zeta}$	m
density of water	ρ_w	Kg/m^3
oil slick total mass	M_s	Kg
thickness of i-th component of oil slick	$\hat{s}_i$	m
density of i-th oil slick component	ρ_i	Kg/m^3
viscosity of i-th oil slick component	μ_i	Kg/ms
surface tension of i-th oil slick component	γ_i^t	Kg/ms^2
velocity of i-th oil slick component	$\tilde{\mathbf{u}}_i$	1/m^2
depth averaged velocity of i-th oil slick component	$\breve{\mathbf{u}}_i$	m/s
outward normal vector to ocean surface	$\hat{\mathbf{n}}$	-
transverse component of wind stress	τ	Kg/ms^2
Eulerian ocean velocity at surface	$\mathbf{u} = (\mathbf{u}, \mathbf{v}, \mathbf{w})$	m/s
slip velocity parameter	C_{xs}	-
pressure, ambient plus dynamic	$\bar{p}' = p_0 + \bar{p}$	Kg/ms^2
wave frequency, peak wave frequency	σ, σ_0	rad/s

depth-averaged transport velocity	$\mathbf{V}(\mathbf{x},t) = \mathbf{v}^c + \mathbf{u}^{st}$	m/s
depth-averaged Eulerian velocity	$\mathbf{v}^c(\mathbf{x},t) = (u^c, v^c)$	m/s
depth-averaged Stokes drift velocity	$\mathbf{u}^{st}(\mathbf{x},t) = (u^{st}, v^{st})$	m/s
tracer dispersion tensor	$\Psi = \Sigma + \Xi \approx D$	m²/s
turbulent Reynolds stress tensor	Σ	m²/s
dispersion caused by averaging	Ξ	m²/s
dispersion due to fluctuations respect to friction velocity	Θ	m²/s
dispersion due to fluctuations respect to Stokes drift	Θ^{st}	m²/s
friction velocity	$\mathbf{v}_*$	m/s
wind speed	v_{wind}	m/s
fractional water content	F_{wc}	-
fraction of evaporated oil from slick	F_{evap}	-
reaction, mass exchange, and other rates of oil slick	E_i^s, R_i, G_i^s	m/s
total mass of subsurface oil	M_c	Kg
concentration of i-th species	C_i	Kg/m³
generic tracer concentration	$B = \bar{B} + B'$	Kg/m³
ocean velocity	$\mathbf{U}(\mathbf{x},z,t) = (U,V,W) = \bar{\mathbf{U}} + \mathbf{U}'$	m/s
tracer molecular diffusion	κ	m²/s
eddy flux tensor	$\mathbf{F}$	Kg/s m²
parcel coordinate	$\mathbf{Z}$	m
Kronecker delta tensor	δ	-
subsurface concentration associated with intermediate time scales	b	m³/m³
wave covariance	e^2	m²
mixing layer thickness, oil mixed layer depth	$P, \xi \approx \min(H(x),P)$	m
wave oil diffusion coefficient	D_w	m²/s
indicator function of oil slick	χ_S	-
absolute and relative group velocity	c_G, C_G	m/s
unidirectional wave spectrum	F	s m²
Current forces: wind, breaking, bottom drag, lateral viscosity	$\mathbf{S}, \mathbf{B}, \mathbf{D}, \mathbf{N}$	m/s²
vortex force	$\mathbf{J}$	m/s²
vorticity, Coriolis constant	$\omega, 2\Omega$	1/s
wind drag parameter	C_D	-
Manning drag parameter	C_M	-
loss term in the action equation	ϵ	Kg/s
energy density	$\mathcal{E}$	Kg/s²
seafloor slope	m	m/m
surf zone extent	L	m
eddy viscosity	K	m²s
subsurface velocity	$u_C(x)$	m/s
subsurface effective bulk oil velocity	$u_e(x) = u_C(x) + D(x)\frac{1}{\xi(x)}\frac{d\xi(x)}{dx}$	m/s
effective thickness of submersed oil	S	m
nearshore bottom drag parameter	d_b	-

References

1. Maltrud, M.; Peacock, S.; Visbeck, M. On the possible long-term fate of oil released in the Deepwater Horizon incident, estimated using ensembles of dye release simulations. *Environ. Res. Lett.* **2010**, *5*, 035301, doi:10.1088/1748-9326/5/3/035301.

2. Liu, Y.; Macfadyen, A.; Ji, Z.G.; Weisberg, R.H.; Huntley, H.S.; Lipphardt, B.L.; Kirwan, A.D. Surface drift predictions of the Deepwater Horizon spill: The Lagrangian perspective. *Environ. Sci. Technol.* **2013**, *46*, 7267–7273.

3. Mariano, A.; Kouralaloua, V.; Srinivasana, A.; Kanga, H.; Halliwell, G.; Ryana, E.; Roffer, M. On the modeling of the 2010 Gulf of Mexico oil spill. *Dyn. Oceans Atmos.* **2011**, *52*, 311–322.

4. Wang, J.; Shen, Y. Modeling oil spills transportation in seas based on unstructured grid, finite-volume, wave-ocean model. *Ocean Model.* **2012**, *35*, 332–344.

5. Tkalich, P.; Chan, E.S. A CFD solution of oil spill problems. *Environ. Model. Softw.* **2006**, *21*, 271–282.

6. Papadimitrakis, J.; Psaltaki, M.; Christolis, M.; Markatos, N.C. Simulating the fate of an oil spill near coastal zones: The case of a spill (from a power plant) at the Greek island of Lesvos. *Environ. Model. Softw.* **2006**, *21*, 170–177.

7. Azevedo, A.; Oliveira, A.; Fortunato, A.B.; Zhang, J.; Baptista, A.M. A cross-scale numerical modeling system for management support of oil spill accidents. *Mar. Pollut. Bull.* **2014**, *80*, 132–147.

8. Seymour, R.J.; Geyer, R. Fates and Effects of Oil Spills. *Annu. Rev. Energy Environ.* **1992**, *17*, 261–283.

9. Feddersen, F. Observations of the surf-zone turbulent dissipation rate. *J. Phys. Oceanogr.* **2012**, *42*, 386–399.

10. Feddersen, F. Scaling surf zone turbulence. *Geophys. Res. Lett.* **2012**, *39*, doi:10.1029/2012GL052970.

11. Fingas, M.F.; Duval, W.; Stevenson, G.B. *The Basics of Oil Spill Cleanup*; Minister of Supply and Services: Quebec, Canada, 1979.

12. Payne, J.R.; Philips, C. *Petroleum Spills in the Marine Environment: Chemistry and Formation of Water-in-Oil Emulsions and Tar Balls*; Lewis Publishers: Chelsea, MI, USA, 1985.

13. Zhang, X.; Hetland, R.; Marta-Almeida, M.; DiMarco, S.F. A numerical investigation of the Mississippi and Atchafalaya freshwater transport, filling and flushing times on the Texas-Louisiana Shelf. *J. Geophys. Res. Oceans* **2012**, *117*, 5705–5723.

14. Zhang, Z.; Hetland, R.; Zhang, X. Wind-modulated buoyancy circulation over the Texas-Lousiana Shelf. *J. Geophys. Res. Oceans* **2013**, *119*, 5705–5723.

15. Warner, J.D.; Geyer, W.R.; Lerczak, J.A. Numerical modeling of an estuary: A comprehensive skill assessment. *J. Geophys. Res. Oceans* **2005**, *110*, doi:10.1029/2004JC002691.

16. Hetland, R.D. Event driven model skill assessment. *Ocean Model.* **2006**, *11*, 214–223.

17. Howlett, E.; Jayko, K. COZOIL (Coastal Zone Oil Spill Model), Improvements and Linkage to a Graphical User Interface. In Proceedings of the Twenty-First Artic and Marine Oilspill Program (AMOP) Technical Seminar, Alberta, Canada, 10–12 June 1998; Volume 21, pp. 541–550.

18. Zhao, L.; Torlapati, J.; Boufadel, M.; King, T.; Robinson, B.; Lee, K. VDROP: A Comprehensive model for droplet formation of oils and gases in liquids—Incorporation of the interfacial tension and droplet viscosity. *Chem. Eng. J.* **2014**, *253*, 93–106.

19. McWilliams, J.C.; Restrepo, J.M.; Lane, E.M. An asymptotic theory for the interaction of waves and currents in coastal waters. *J. Fluid Mech.* **2004**, *511*, 135–178.

20. Lehr, W.J. Review of modeling procedures for oil spill weathering behavior. In *Advances in Ecological Models*; Brebbia, C.A., Ed.; WIT Press: Southampton, UK, 2001; Chapter 3.

21. McWilliams, J.C.; Restrepo, J.M. The wave-driven ocean circulation. *J. Phys. Oceanogr.* **1999**, *29*, 2523–2540.

22. Restrepo, J.M.; Ramírez, J.M.; McWilliams, J.C.; Banner, M. Multiscale momentum flux and diffusion due to whitecapping in wave–current interactions. *J. Phys. Oceanogr.* **2011**, *41*, 837–856.

23. Garrett, C. Turbulent dispersion in the ocean. *Progress Oceanogr.* **2006**, *70*, 113–125.

24. Kim, D.H.; Lynett, P.J. Turbulent mixing and passive scalar transport in shallow flows. *Phys. Fluids* **2011**, *23*, 016603, doi:10.1063/1.3531716.

25. Atlas, R.M.; Hazen, T.C. Oil biodegradation and bioremediation: A tale of the two torst spills in U.S. history. *Environ. Sci. Technol.* **2011**, *45*, 6709–6715.

26. Moghimi, S.; Restrepo, J.M.; Venkataramani, S. Mass conservation modeling in an oil fate model for shallow waters. *J. Mar. Sci. Eng.* 2016, submitted for publication.

27. Lasheras, J.C.; Eastwood, C.; Martínez-Bazán, C.; Montañés, J. A review of statistical models for the break-up of an immiscible fluid immersed into a fully turbulent flow. *Int. J. Multiph. Flows* **2002**, *28*, 247–278.

28. Boehm, P.; Flest, D.L.; Mackay, D.; Paterson, S. Physical-chemical weathering of petroleum hydrocarbons from the Ixtoc I blowout: Chemical measurements and a weathering model. *Environ. Sci. Technol.* **1982**, *16*, 498–504.

29. Cuesta, I.; Grau, F.X.; Giralt, F. Numerical simulation of oil spills in a generalized domain. *Oil Chem. Pollut.* **1990**, *7*, 143–159.

30. Quinn, M.F.; Marron, K.; Paten, B.; Abu-Tabanja, R.; Al-Bahrani, H. Modeling of the ageing of crude oils. *Oil Chem. Pollut.* **1990**, *7*, 119–128.

31. Buchanan, I.; Hurford, N. Methods for predicting physical changes in oil spilt at sea. *Oil Chem. Pollut.* **1988**, *4*, 311–328.

32. Sebastiao, P.; Soares, C.G. Modeling the fate of oil spills at sea. *Spill Sci. Technol. Bull.* **1995**, *2*, 121–131.

33. Reed, M.; Gundlach, E.; Kana, T. A coastal zone poil spill model: Development and sensitivity studies. *Oil Chem. Pollut.* **1989**, *5*, 411–449.

34. Mooney, M. The viscosity of a concentrated suspection of spherical particles. *J. Colloidal Sci.* **1951**, *10*, 162–170.

35. Mackay, D.; Shiu, W.Y.; Hossain, K.; Stiver, W.; McCurdy, D.; Paterson, S.; Tebeau, P. *Development and Calibration of an Oil Spill Behaviour Model*; CG-D-27-83; US Coast Guard Research and Development Center: Washington, DC, USA, 1982.

36. Xie, H.; Yappa, P.D.; Nakata, K. Modeling emulsification after an oil spill in the sea. *J. Mar. Syst.* **2007**, *68*, 489–506.

37. Stiver, W.; Mackay, D. Evaporation rate of spills of hydrocarbons and petroleum mixtures. *Environ. Sci. Technol.* **1984**, *18*, 834–840.

38. Stiver, W.; Shiu, W.Y.; Mackay, D. Evaporation times and rates of specific hydrocarbons in oil spills. *Environ. Sci. Technol.* **1989**, *23*, 101–105.

39. Mackay, D.; Matsugu, R.S. Evaporation Rates of liquid hydrocarbon spills on land and water. *Can. J. Chem. Eng.* **1973**, *51*, 434–439.

40. Bobra, M. *A Study of the Evaporation of Petroleum Oils*; MR EE-135; Environment Canada, Environmental Protection Directorate, River Road Environmental Technology Centre: 1992.

41. Smith, C.L.; McIntyre, W.G. Initial Ageing of Fuel Oils on Sea Water. In Proceedings of the Joint Conference on Prevention and Control of Oil Spills, Washington, DC, USA, 15–17 June 1971; pp. 457–461.

42. Fingas, M. A literature review of the physics and predictive modeling of oil spill evaporation. *J. Hazard. Mater.* **1995**, *42*, 157–175.

43. Payne, J.R.; Phillips, C.W. Photochemistry of oil in water. *Environ. Sci. Technol.* **1985**, *19*, 569–579.

44. Spaulding, M.L. A state-of-the-art review of oil spill trajectory and fate modeling. *Oil Chem. Pollut.* **1988**, *4*, 39–55.

45. Webb, A.; Fox-Kemper, B. Wave spectral moments and Stokes drift estimation. *Ocean Model.* **2011**, *40*, 273–288.

46. Church, J.C.; Thornton, E.B. Effects of breaking wave induced turbulence within a longshore current model. *Coast. Eng.* **1993**, *20*, 1–28.

47. Thornton, E.B.; Guza, R.T. Transformation of wave height distribution. *J. Geophys. Res.* **1983**, *88* (C10), 5925–5938.

48. Venkataramani, S.C.; Venkataramani, R.; Restrepo, J.M. Stochastic parametrization of weathering processes. *Phys. Rev. E* 2016, submitted for publication.

49. Restrepo, J.M.; Venkataramani, S.C.; Dawson, C. Nearshore sticky waters. *Ocean Model.* **2014**, *80*, 49–58.

50. Restrepo, J.; Venkataramani, S.; Balci, N. Stochastic longshore dynamics equations. *Ocean Model.* **2015**, under review.

51. Longuet-Higgins, M.S. Longshore currents generated by obliquely incident sea waves: 1 & 2. *J. Geophys. Res.* **1970**, *75*, 6778–6801.

52. Svendsen, I.A.; Putrevu, U. Surf-zone hydrodynamics. *Adv. Coast. Ocean Eng.* **1995**, *2*, 1–78.

53. Thornton, E.B.; Guza, R.T. Surf zone longshore currents and random waves: Field data and models. *J. Phys. Oceanogr.* **1986**, *16*, 1165–1178.

54. Guannel, G.; Özkan-Haller, H.T. Formulation of the undertow using linear wave theory. *Phys. Fluids* **2014**, *26*, 056604, doi:10.1063/1.4872160.

55. McWilliams, J.C.; Sullivan, P.P.; Moeng, C.H. Langmuir turbulence in the ocean. *J. Fluid Mech.* **1997**, *334*, 1–30.

56. Johnson, D.R. *Ocean Surface Current Climatology in the Northern Gulf of Mexico*; Gulf Coast Research Laboratory: Ocean Springs, MS, USA, 2008.

57. Bathymetric Data Viewer. Available online: http://maps.ngdc.noaa.gov/viewers/bathymetry/ (accessed on 2 December 2015).

58. McWilliams, J.C. A perspective on submesoscale geophysical turbulence. In *IUTAM Symposium on Turbulence in the Atmosphere and Oceans*; Dritschel, D., Ed.; Springer: Berlin, Germany, 2010; pp. 131–141.

59. Capet, X.; McWilliams, J.C.; Molemaker, M.J.; Shchepetkin, A. Mesoscale to submesoscale transition in the California Current System. Part I: Flow structure, eddy flux, and observational tests. *J. Phys. Oceanogr.* **2008**, *38*, 29–43.

60. Shchepetkin, A.F.; McWilliams, J. The regional oceanic modeling system (ROMS): A split-explicit, free-surface, topography-following-coordinate oceanic model. *Ocean Model.* **2005**, *9*, 347–404.

Journal of
*Marine Science
and Engineering*

MDPI

Article

Consensus Ecological Risk Assessment of Potential Transportation-related Bakken and Dilbit Crude Oil Spills in the Delaware Bay Watershed, USA

Ann Hayward Walker [1,*], Clay Stern [2], Debra Scholz [3], Eric Nielsen [4], Frank Csulak [5] and Rich Gaudiosi [6]

[1]   SEA Consulting Group, Cape Charles, VA 23310, USA
[2]   US Fish and Wildlife Service (USFWS), Galloway, NJ 08205, USA; clay_stern@fws.gov
[3]   SEA Consulting Group, Charleston, SC 29412, USA; dscholz@seaconsulting.com
[4]   US Coast Guard (USCG), Sector Delaware Bay, Philadelphia, PA 19147, USA; Eric.D.Nielsen@uscg.mil
[5]   National Oceanic and Atmospheric Administration (NOAA), Emergency Response Division, Highlands, NJ 07732, USA; frank.csulak@noaa.gov
[6]   Delaware Bay and River Cooperative, Inc. (DBRC), Lewes, DE 19958, USA; Rich_Gaudiosi@dbrcinc.org
*   Correspondance: ahwalker@seaconsulting.com; Tel.: +1-757-331-1787

Academic Editor: Merv Fingas
Received: 16 January 2016; Accepted: 17 February 2016; Published: 7 March 2016

Abstract: Unconventionally-produced crude oils, *i.e.*, Bakken oil and bitumen diluted for transport and known as dilbit, have become prominent components of the North American petroleum industry. Spills of these oils have occurred during transport from production areas to refineries via pipeline, rail, and barge. Some of their physical and chemical properties are distinct and present new challenges in mitigating spill impacts on people and the environment. This paper describes the adaptation of a qualitative risk assessment process to improve spill preparedness and response decisions for these oils when transported in an estuarine area. The application of this collaborative, interdisciplinary process drew upon a literature review, the local knowledge and experience of a broad set of decision makers, practitioners, and technical experts who developed consensus-based recommendations aimed at improving response to spills of these oils. Two emphasized components of this consensus ecological risk assessment (CERA) concerned risks: (1) to human health and safety and (2) from spilled oil and the associated response actions on endangered species. Participants in the process defined levels of concern associated with Bakken and dilbit oils relative to a set of response actions in freshwater, brackish and saltwater habitats and on resources at risk.

Keywords: Bakken; bitumen; dilbit; risk assessment; oil spill; response; preparedness; endangered species; threatened species; consensus; human health

1. Introduction

Crude oils produced by unconventional methods in North America have become marketable resources used to meet energy demand in the United States (U.S.) and elsewhere. Unconventional crude oils derive from two sources. In the U.S., hydraulic fracturing technologies have been widely applied to extract oil from shale formations or other typically inaccessible, low-permeability rocks. In Canada, petroleum products have been extracted from "oil sands" or "tar sands" [1]. An oil shale formation is a fine-grained sedimentary rock containing a solid material (kerogen) that converts to liquid oil when heated. Oil shale deposits globally occur in 37 countries; the largest and highest quality oil shale deposits are located in sparsely populated areas of Colorado, Utah and Wyoming [2]. Estimated volumes of these oil reserves have increased steadily over the last six years to 39.9 billion barrels in 2014, which represents an increase of 9.3% over the previous year [3].

The crude oils produced from oil shale formations and tar sands fields are being transported by pipeline, rail, barge and tanker to refineries and market centers (Figure 1). Crude oil with high sulfur content is referred to as "sour", while oil with low sulfur content is considered "sweet oil". Bakken oil is classified as a light, sweet crude oil, which is transported from shale formations to market by rail cars [4], often in as many as 100 cars in a single shipment, referred to as a unit train. Bitumen is a heavy, sour oil derived from oil sand formations in Alberta, Canada. It is a mixture of heavy oil, sand, clay and water; then separated from the sand and water in a centrifuge. Bitumen is mixed with about 30% of diluents in order to decrease viscosity and facilitate flow during transportation via pipeline; it is known as diluted bitumen, or "dilbit". When transported by rail, bitumen is diluted about 15% with a diluent and then known as "railbit" [5].

Figure 1. Major rail transportation arteries for oil across North America [4].

Recent transportation-related incidents have resulted in spills of these unconventional crude oils (Table 1). Bakken and dilbit oils exhibit some properties which present distinct issues for emergency responders. Bakken oils are highly volatile and soluble in water. Accidents involving unit trains that are transporting Bakken oil have led to serious fires and loss of life. Dilbit oil is heavy crude oil. Compared to medium or light crude oils, it is characterized by exceptionally high density, viscosity, and adhesion properties from the bitumen component of the diluted bitumen. For both these crude oils, these properties affect weathering behavior (physical and chemical changes of spilled oil from its initial release over time) in the environment, especially when dilbit is spilled into waterways.

Table 1. Examples of recent transportation-related incidents involving spills of dilbit and Bakken crude oil [6–14].

Description	Volume	Comments
Dilbit Oil		
June 2010—Marshall, Michigan, USA	Enbridge Energy Partners Limited Liability Partnership's (Enbridge) 30-inch pipeline ruptured releasing its contents, *i.e.*, dilbit (report to the National Response Center was 19,500 bbls) into a culvert leading to the Tallmadge Creek, a tributary of the Kalamazoo River. The oil sank to the river's bottom and collateral damage resulted from recovery tactics.	• The Kalamazoo River is bordered by marshland and developed properties for the approximate 30-mile stretch of the response site • Variety of tactics to collect the oil: spraying the sediments with water, dragging chains through the sediments, agitating sediments by hand with a rake, and driving back and forth with a tracked vehicle to stir up the sediments and release oil trapped in the mud
March 2013—Mayflower, Arkansas, USA	ExxonMobil's 20-inch "Pegasus Pipeline" ruptured near Mayflower, Arkansas. Approximately 5000 bbls of dilbit (Canadian Wabasca heavy crude oil from the Athabasca oil sands) spilled into the surrounding area and flowed into Lake Conway.	• Approximately 62 homes evacuated • Wetland vegetation, waterfowl and various other wildlife were impacted • Personnel on Site (250 in Command Post/440 in Field) • 663 ExxonMobil personnel and contractors • 23 Vacuum Trucks • 85 Frac tanks • 9593 ft. Hard Boom and 241,290 ft. of Soft Boom
Bakken Oil		
April 2014—Lynchburg, Virginia, USA	A CSX train carrying Bakken crude oil in a 105-car train, jumped the rails causing 13-unit cars to derail and some were damaged. The derailment sparked a large fire that forced the evacuation of six city blocks; 3 cars submerged in the James River, and 30,000 gals of oil were released.	• No fatalities or injuries • Fire permitted to burn • Soil and vegetation was coated with crude oil • 17-mile oil slick in the James River
February 2014—Mississippi River, Louisiana, USA	The Tank Barge E2MS 303 collided with the towboat Lindsay Ann Erickson on the Lower Mississippi River (between Baton Rouge and New Orleans), causing a spill of approximately 750 bbls (31,500 gals) of Bakken oil.	• 65-mile closure of the Lower Mississippi River for 2 days • Total oil recovered: 95 gals (2.3 bbls) • Reports of high concentrations of benzene vapors during lightering operations • Approximately 150 personnel responding to this incident including federal, state, local, and industry representatives. • No reports of oiled or injured wildlife

Table 1. *Cont.*

Description	Volume	Comments
November 2013—Aliceville, Alabama, USA	90-car train was crossing a timber trestle above a wetland near Aliceville late Thursday night when approximately 25 rail cars and two locomotives derailed, spilling Bakken crude oil into the surrounding wetlands and igniting a fire that was still burning Saturday. Each of the 90 cars was carrying 30,000 gals of oil; 630,000 gals were either spilled or burned.	• 21 cars were in marsh habitat • Significant fire permitted to burn • No fatalities or injuries • Response hampered by lack of access in remote area • Significant impact to wetlands
July 2013—Lac-Megantic, Quebec, Canada	Runaway train derailed (insufficient hand breaks) with 63 rail cars (30,000 gals each). Approximately 1.7 million gals of Bakken oil either burned or was released, with an estimated 26,000 gals into the Chaudière River.	• Massive fire in town center • 47 residents were killed • Over 70 buildings destroyed • River restricted to non-drinking water status

To improve response capabilities to potential incidents, emergency response decision makers throughout the USA, Canada, and the European Union have studied these past oil spills and prepared recommendations for their respective jurisdictions [15–29]. In the USA, the USA Coast Guard (USCG) is the lead federal authority for oil spill response decisions in coastal areas; the US Environmental Protection Agency (EPA) is the lead federal authority for inland areas. USCG and EPA lead response officials are the designated Federal On-scene Coordinators (FOSCs). Other government agencies and stakeholders, including potentially responsible parties, have roles in oil spill preparedness and response as described in Subparts B and C of the National Oil and Hazardous Substances Pollution Contingency Plan, referred to as the NCP [30]. The increase in production of these unconventionally-derived oils has resulted in changes in the transportation patterns and in oil spill risks in the coastal zone [31], to which the USCG has turned its attention. The USCG is a multi-mission agency, charged with enforcing various laws and regulations and protection of maritime economy, the environment and ultimately providing for the well-being, general safety, security, and interests of the citizens of the United States. In this regard, the USCG must consider and manage potential risks to transportation safety and the marine environment from pollution.

A variety of risk assessment approaches are available; for example the NEREIDs [32] supported by cooperation of marine research centers in Greece, Cyprus and the United Kingdom, along with a multinational incident report database advances cross-border civil protection and marine pollution cooperation for direct response to natural and man-made disasters. Modeling of oil spills in confined maritime basins using bathymetric and geomorphic data has been used to identify risks to vulnerable and sensitive coastlines [28,29,33]. In 1998, Aurand [34] proposed adapting EPA's Ecological Risk Assessment (ERA) approach, which was first developed in 1992 to assess ecological effects caused by human activities, for use in oil spill planning [35]. The goal was to facilitate resolution of disagreements over the potential risks and benefits of alternative response technologies, *i.e.*, dispersants and *in-situ* burning, compared to traditional mechanical containment and recovery actions. In 1998, the concept was put into practice for the first time in a set of meetings in the Puget Sound area [36]. This initial effort clearly demonstrated the need for, and challenges with, reconciling concerns among environmental response professionals with diverse expertise and responsibilities about which response actions would have the lowest collateral damage and would reduce the risks of impacts from spilled oil. The situation was more complex than just committing to the use of ecological risk assessment methods to overcome skepticism and concerns about the environmental risks of using dispersants and *in-situ* burning in addition to mechanical recovery during oil spill response. Based on the Puget Sound ERA, the process was revised to place more emphasis on facilitated risk communication and qualitative risk assessment, and less emphasis on attempts to quantify the details related to a hypothetical event. The revised process, which emphasizes reaching agreement among the risk management team, stakeholders, and interested parties, has been referred to as a Consensus Ecological Risk Assessment (CERA).

The USCG developed a guidebook for conducting CERAs [37] and over a dozen have been carried out. Briefly, the CERA requires a collaborative, multi-disciplinary exchange and application of scientific findings from currently available literature in addition to local knowledge, and direct experience of risk managers (spill decision makers), resource agency managers and scientists (risk assessors), and other interested parties. The process involves a comparison of the anticipated impacts of spilled oil to the potential mitigation of impacts by various response actions with their associated collateral damages. The objective is to identify those response actions which are most likely to mitigate overall spill risks on people and the environment, that is, improve the outcome over letting the oil attenuate naturally.

Building upon the process used in previous CERAs, the USCG FOSC for Sector Delaware Bay initiated a CERA in 2014 with the aim of improving preparedness and response to potential transportation-related spills of Bakken and dilbit crude oils in the Delaware Bay watershed. The changing energy landscape and associated risks to ecological resources necessitates a whole-of-community approach to improving preparedness and response to spills of these crude oils.

The focus of this paper is the adaptation of the USCG CERA to evaluate a different set of response actions than previous CERAs for spills involving Bakken and dilbit, which have become part of the USCG's evolving responsibilities in the recent domestic energy renaissance, particularly in the Delaware Bay Watershed. This work incorporates a new emphasis to address potential risks to US federal- and state-listed threatened and endangered (T/E) species. An overview of the CERA process is presented in the Methods Section. The Discussion Section describes noteworthy adaptations by the Project Committee of important components in the process, *i.e.*, resources of concern, response actions, conceptual model, levels of concern, that were used to characterize the potential risks to people and the environment from transportation-related spills of Bakken and dilbit oils. The project report contains a detailed description of this work [38].

2. Methods

The USCG Guidebook describes details for conducting a CERA for a marine oil spill [37]. Applications of earlier CERAs are discussed elsewhere [39–41] and are similar to those implemented in the NEREIDs [32] project in Europe. This CERA was consistent with the 12 activities, carried out in four phases, as described in the USCG Guidebook (Figure 2). The CERA process was guided by a Project Committee, comprised of agency representatives in the study area from the USCG, EPA, National Oceanic and Atmospheric Administration (NOAA), U.S. Fish and Wildlife Service (USFWS), the States of Delaware (DE), Pennsylvania (PA), and New Jersey (NJ), an Oil Spill Response Organization (OSRO) and the consultancy firm responsible for carrying out the work. Many previous CERAs that were conducted using the USCG Guidebook were reviewed as background for this project, along with a broad review of the relevant literature, access to which was provided to all participants on a SharePoint site. A USCG objective in this CERA was also to consider risks to T/E species in accordance with USCG headquarters guidance [42], and pursuant to Endangered Species Act (16 U.S.C. 1531–1544, 87 Stat. 884) Section 7(a) (1–2) and Essential Fish Habitat under the Magnuson-Stevens Fisheries Conservation and Management Act.

PHASE 1: Problem Definition/Formulation

1. Assemble the Project Committee
2. Develop the scenario(s)
3. Estimate the transport, fate of oil, and exposure potential
4. Define response actions for consideration
5. Define resources of concern

PHASE 2: Conceptual Model/Analysis Plan (All participants, Workshop 1)

6. Consider important relationships
7. Thresholds of sensitivity to oil

PHASE 3: Analysis and Risk Characterization (All participants, Workshop 2)

8. Determine levels of concern about effects
9. Evaluate relative risk for oil only *vs.* various response actions
10. Define limits of the analysis

PHASE 4. Document and Apply (Project Committee)

Figure 2. Consensus Ecological Risk Assessment process for oil spills.

2.1. Risk Analysis and Characterization

The risk analysis and characterization was conducted during two, two-day interactive workshops, separated by about two weeks. The scope of this CERA was considerably more complex than that in previous CERAs. For this reason, the Project Committee developed in advance and presented many of the outputs of activities 2–7 (Figure 2) to workshop participants for their consideration and finalization. A total of 88 participants provided their input. They represented the following groups of stakeholders: USCG; EPA; US Department of the Interior; USFWS; NOAA; Federal Emergency Management Agency; Agency for Toxic Substances and Diseases Registry; the States of PA, NJ, and DE; City of Philadelphia; Delaware County, PA; Fire Departments of Philadelphia and Eddystone, PA; academia; non-governmental organizations (NGOs); oil spill responders; and rail and petroleum industries. Participants were assigned to scenario-specific workgroups to reach consensus about the relative risk characterizations for five pre-determined scenarios. They also met in plenary sessions at the beginning and end of each day for group presentations, briefings, and discussion of their respective findings. Drawing on the collective experience and knowledge of a cross-section of the local oil spill response community, the participants qualitatively evaluated the adverse ecological impacts of spilled oil and spilled oil plus an evaluated response action to predict the severity and duration of adverse impact to natural resources in a given scenario.

2.2. Study Area

The geographic area of concern for this CERA encompassed the coastal zone, a portion of the USCG area of responsibility in the Delaware Bay Estuary in the mid-Atlantic region of the US (Figure 3), that included marine (coastal bay), brackish (coastal river) and freshwater habitats (inland river). The Delaware Bay region is a nationally and internationally important natural resource and is a critical component of U.S. energy independence which ultimately contributes to national security. According to the U.S. Energy Information Administration (EIA), crude by rail movements of Bakken and dilbit to the mid-Atlantic region have increased from 1000 barrels per day (BPD) in 2010, to 800,000 BPD in 2014 (Figures 4 and 5), including the Delaware Bay region in 2014.

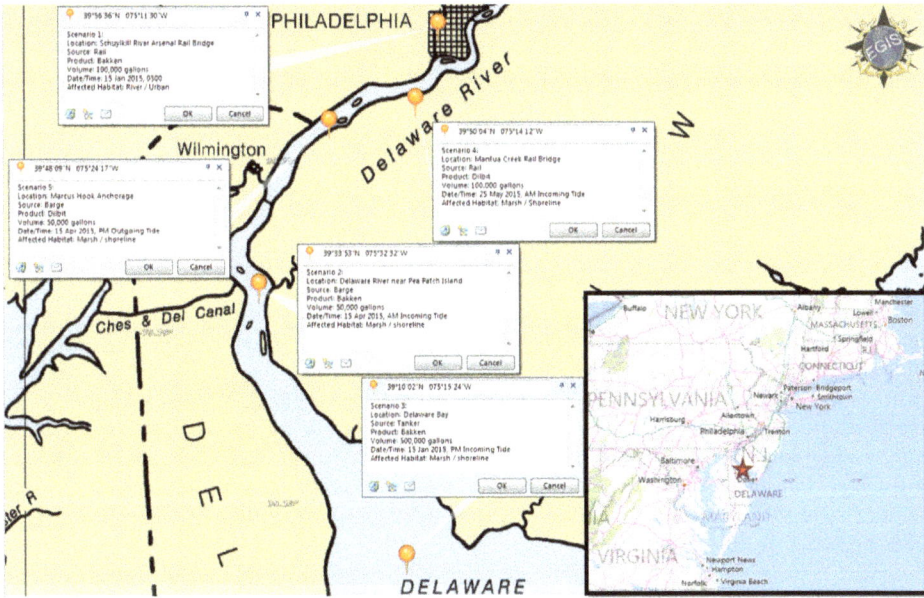

Figure 3. Locations of the five scenarios. Inset shows the location of the study area—the Delaware Bay estuary in the mid-Atlantic region of the US.

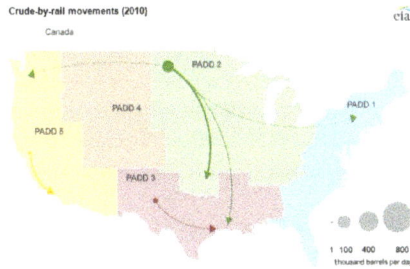

Figure 4. Crude-By-Rail Movements of Bakken and dilbit in 2010, tracked by Petroleum Administration for Defense District (PADD) regions [43].

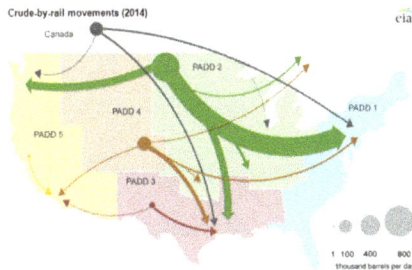

Figure 5. Crude-By-Rail Movements of Bakken and dilbit in 2014, tracked by Petroleum Administration for Defense District (PADD) regions [44].

The Delaware Bay Estuary is regarded internationally as ecologically valuable, designated as a Wetland of International Importance by the Ramsar Convention. In May 1986, Delaware Bay was recognized as the first Western Hemisphere Shorebird Reserve Network (WHSRN) Site of Hemispheric Importance for migrating shorebirds, and is considered the most important spring stopover in the eastern USA for several species of shorebirds [45]. Using natural capitalization valuation methods, estimates of the value of natural goods and ecological services provide by the Delaware Bay Estuary watershed exceeded $10 billion per year [46].

2.3. Transportation Scenarios

The Project Committee considered locally-probabilistic scenarios to define the risk situation as the basis for subsequent activities. In this case, five scenarios reflecting representative transportation patterns, risks of interest to local and regional stakeholders, and response considerations, e.g., response times and efficacy of response actions were defined. Variables between scenarios included: waterway type, (urban (river) and more rural (bay and creek) environments; transportation mode (rail, barge and tanker) seasonal differences (winter and spring) to address the differences in response actions and sensitivity of resources of concern. The locations of the five scenarios are shown in Figure 3. Scenarios 1–3 involved spills of Bakken oil from rail cars (Scenario 1), a barge (Scenario 2), and a tanker (Scenario 3); Scenarios 4–5 involved spills of dilbit oil from rail cars (Scenario 4) and a barge (Scenario 5).

2.4. Threatened and Endangered Species

During an oil spill, as matter of federal law, the FOSC is required to consult with the appropriate federal resource agencies to ensure that the response actions taken are not likely to jeopardize the continued existence of any federally-listed threatened or endangered species or result in the destruction or adverse modification of habitat of such species. In the USA, animal or plant species of conservation concern may be listed as threatened or endangered under the USA Endangered Species Act (federally-listed species). A federal listing means that a species is in decline throughout its entire range which may encompass several other states or nations. Individual states have the authority to list a species of conservation concern (state-listed species), but this listing only addresses species status within that state. As such, species may be listed as state threatened or endangered and not be federally listed. Therefore, for the conservation benefit of state-listed species, the participants came to consensus early in a process that state-listed species would be given extra consideration similar to that afforded by law to federally-listed species throughout the CERA. In doing so, participants were galvanized towards using their individual and collective authorities in carrying out the CERA to proactively further the conservation of threatened and endangered species independent of whether the species was federally or state listed. For the purposes of this CERA, use of the term "T/E species" is inclusive of federal and state-listed species. Example T/E plants and animals in the study area included among others:

- Mammals—northern long-eared bat,
- Birds—red knot and least bittern (shorebirds),
- Amphibians—southern leopard frog,
- Fish—Atlantic sturgeon, short-nose sturgeon, hickory shad, banded sunfish,
- Reptiles—eastern redbelly turtle, and
- Plants—long-lobed arrowhead, wild rice, and seabeach amaranth.

3. Results

Participants acting in the roles of risk managers, risk assessors, or other interested and affected stakeholders, characterized the potential ecological and human health and safety risks of the five scenarios. The findings in this CERA summarize the workgroup-specific consensus for each of five scenarios, rather than consensus of all participants for all five scenarios. When spilled, Bakken and dilbit behave differently. Moreover, the behavior of dilbit oils can vary depending upon the type and percentage of added diluents. For Bakken oils, participants agreed that the primary initial strategy is to safely mitigate risks from flammable vapors. For dilbit oils, participants determined that it is imperative for containment and skimming operations to be implemented immediately to recover the oil before it spreads, weathers, begins to pick up any sediments in the water, and possibly would begin to sink. However, in developing response strategies during the initial hours of a spill, responders must also consider health and safety hazards of the oils, *i.e.*, evaporation of light ends, release of hydrogen sulfide gas, and flammability, especially with regard to ignition sources, e.g., boat engines. In this CERA, participants assigned higher levels of concern about the ecological risks associated with a spill of dilbit oil compared to a spill of Bakken crude oil.

The adaptation of the CERA process, to provide a structured way to qualitatively assess the potential risk from transportation-related spills of Bakken and dilbit oil in the coastal zone of the Delaware Bay estuary, was successful. It enabled the interdisciplinary collaboration of risk managers (spill decision makers and resource managers), assessors (agency and academic scientists, responders) NGOs and other stakeholders to reach consensus about the relative risks of the spilled oil compared to a set of response actions on resources of concern, including T/E species. This consensus represents guidance for the lead federal official, *i.e.*, USCG FOSC, about response actions which should be taken to mitigate pollution risks from these oils. Actions to mitigate the public health and safety risks in the early stages of the emergency, especially from rail incidents near populated areas, would remain the responsibility of first responders, e.g., fire fighters.

The CERA process, as conducted, was interdisciplinary. It promoted and enabled the transparent integration of information, data, techniques, tools, perspectives, concepts, and/or theories from multiple disciplines or bodies of specialized knowledge to solve problems such as mitigating human health and ecological risks from transportation-related spills of two unconventional crude oils, whose solutions are beyond the scope of a single discipline or area of research practice. As such, this CERA was consistent with the findings of a study that established a consensus of building principles for a landscape approach to reconciling conservation with other competing land uses [47].

The CERA process was designed as a planning and educational tool to enable thoughtful, comprehensive assessment of potential risks. It is not designed, nor intended, to be implemented real-time during response. Notwithstanding, the knowledge and experience gained by the participants facilitates real-time decision-making among risk and resource managers if an actual spill occurs in the same geographic area because the detailed reasoning process of potential risks and benefits of response actions remains relevant, even when the specific details of the situation are different. Further, the process gives equal opportunity to all participants to provide input into the risk characterization. This aspect of the CERA cultivates and strengthens mutually-beneficial working relationships and, when consensus is reached, builds trust that the chosen response actions are the appropriate ones to mitigate ecological and human health risks.

The process recognized that the many values and services provided by the resources of concern are, to some degree, measured by the judgement of resource managers, and other stakeholders, not as simply by an instrument reading or other singular metric. This qualitative risk assessment considered conservation of the Delaware Bay that allows multiple uses of the waterways as legitimate, *i.e.*, that conservation of the creeks, river, and bay is important for transportation, commerce and recreation, and as habitat for wildlife.

Participants also concluded that there were important gaps in the available information to resolve their questions and concerns around the use of firefighting foam to mitigate fire risks for public health and safety that could potentially present substantive or even unacceptable ecological risks. They advocated that the response community, working with industry, develop recommendations for a holistic "concept of operations" approach for dilbit oil spills that is, define a priority sequence of response actions to implement near the spill source. For example, pre-spill planning should focus on improving the effectiveness of oil containment and recovery in the early stages of a spill. The best option to minimize the ecological risk of dilbit oils using currently available technology is to deliver containment and recovery equipment quickly to the scene to limit the geographic spread of oil and recover the majority of spilled oil before it becomes sufficiently weathered to sink, thereby impeding detection and recovery. In a creek, skimming would be the primary strategy to recover the oil and limit the extent of its movement and contamination, *i.e.*, before the oil could move into the river and become much more difficult to recover.

4. Discussion

The consensus-building approach clearly enabled the diverse members of the Project Committee to broadly consider and work toward a common goal. This collaboration was built on mutual respect for one another's respective specialties, knowledge and responsibilities. In working through this transparent process, participants developed trust in one another's judgements. While the context of this project was an emergency caused by an oil spill, the same stakeholders would be involved in mitigating human health and environmental hazards associated with other emergencies, e.g., extreme weather or terrorism events. In this regard, this CERA added value to all-hazard preparedness in the Delaware Bay Watershed.

Working together, the Project Committee developed new CERA components for assessing risks from transportation-related Bakken and dilbit spills, described below. While the percentage of diluents added to bitumen varies for rail and pipeline transportation, the weathering behavior is the same and therefore this CERA considers dilbit as generally representative of diluted bitumen.

4.1. Resources of Concern

The habitats for the scenarios spanned freshwater (Scenarios 1 and 4), brackish water (Scenarios 2 and 3), and saltwater environments (Scenario 5). Resources of concern were identified for these environments using the structure presented in Table 2. Generally, the categories identified in Table 2 provided the participants a common framework that would be applied to each scenario. The list of potential resources of concern and at risk in the five scenarios were grouped according to habitat, sub-habitat, and resource categories. The presence of T/E species varied on a scenario-specific basis.

Table 2. Organization of resources of concern and at risk from Bakken and dilbit oil spills in marine, brackish water, and freshwater environments in a temperate climate in the mid-Atlantic coast of the USA, encompassing Delaware Bay.

Habitats:
• Artificial shorelines
• Natural terrestrial shorelines
• Intertidal shorelines (including surface waters 0–1 meter)
• Mid-water (0–2 meters from the surface, but above the bottom 2 meters)
• Benthic (bottom + 2 meters)
Sub-habitats:
• Bulkheads, riprap, man-made structures, pavement
• Vegetated, sand, gravel
• Marsh, swamp, tidal flats
• Water column
• Seabed
• Socio-economic resources
• Human health receptors

Table 2. *Cont.*

Categories of Ecological Resources of Concern:
• Mammals (aquatic and non-aquatic dependent)
• Birds (aquatic and non-aquatic dependent)
• Reptiles and amphibians (aquatic and non-aquatic dependent)
• Macro-invertebrates
• Aquatic vertebrates
• T/E species—Animals
• T/E species—PlantsPlants (submerged and floating aquatic vegetation)
• Fishing (commercial and recreational)
• Water intakes (surface and mid-water)

Categories of Socio-economic Resources of Concern:
• Workers
• Residential community
• Sensitive receptors
• Commercial community
• Industrial community
• Transportation community

Because these oils present public safety risks when initially spilled, human health and safety risks also needed to be considered, especially in populated areas, *i.e.*, urban areas and mass transportation corridors. Risks to human health were considered for groups of socio-economic resources of concern, including workers as well as the general public, also in Table 2.

4.2. Response Actions

The Project Committee considered practical categories of currently available response actions that could be implemented during response to spills involving these oils, as well as their logistics limitations, and effectiveness considerations. The potential collateral damages that could result from implementing these response actions (e.g., physical trauma to organisms and habitats from shoreline cleanup, underwater recovery or physical contact methods of oil detection) were considered in the development of the conceptual models for each scenario, which identified the ways in which resources of concern could be exposed to potential hazards associated with the oil and response actions.

Response to spills of these crude oils involves two weathering timeframes: the initial flammability phase when light ends of the oils are present and fires could occur, during which the deployment of traditional spill response options would be pre-empted by first responder (fire fighter) actions; and the second, longer-term phase of responding to the oil on-water. For purposes of this CERA, pollution responders could become actively engaged in the initial 4–6 h after first responders (e.g., fire fighters) would have arrived on scene and might still be dealing with flammability risks. Recent incidents have resulted in significant fires involving Bakken oil, which has been known to re-ignite. Flammability is also a concern with freshly-spilled dilbit oil. During this emergency phase, public safety actions would take precedence over pollution response actions. The behavior of the oil will begin to change due to weathering after the initial 4–6 h.

Next, the CERA considers the weathered oil behavior approximately four to seven days after the emergency phase during which oil could still be found on the water surface and be recoverable using traditional pollution response techniques. Toward the end of this timeframe, the residual bitumen component of dilbit oil would likely begin to pick up sediment in the water column and sink below the water surface, either in the water column or settle on the bottom. The time scales associated with response actions are not absolute; rather they represent a range of hours and days that generally align with important oil weathering and behavior changes that ultimately would influence decisions about potential response actions.

For both of these crude oils, oil recovery on-water is difficult after the oil weathers. Bakken oil is a light crude oil; following rapid evaporation of light ends, the remaining components naturally disperse into the water column, making recovery from the water generally impractical. Dilbit oil, on the other hand, is comprised of heavy oil tar sands mixed with diluents to facilitate its transportation. When initially spilled, the light fractions in the diluted bitumen begin to evaporate. After a few days, residual heavier components may be exposed to sediments in the water column and no longer float, making it more difficult to track and recover.

The following categories of response actions were used in assessing the risks associated with responding to Bakken and dilbit oil spills in the five transportation-related scenarios:

1. Natural attenuation with monitoring (NAM)
2. Fire—Let burn and controlled burn (both *in-situ*)
3. Fire—Extinguishing agent and methods
4. No Fire—Vapor suppression
5. No Fire—Oil spread control (on-land, on-water, and underwater)
6. No Fire—On-water recovery and underwater recovery
7. No Fire—Resource protection (on-water and on-land)
8. No Fire—Shoreline clean-up
9. No Fire—Oil detection/mapping (physical-contact methods)
10. No Fire—Oil detection/mapping (remotely-observed methods)

The overall list was categorized to better align with the response categories of previous ERAs and to facilitate evaluating their risks. These general categories provided a common framework for the workgroup participants since specific response actions could vary among the five scenarios, e.g., some on land, some on water, and presence or absence of ice. The following section defines each possible response action and lists some points regarding logistical considerations, limitations, and considerations that influence its effectiveness in mitigating threats presented by spilled Bakken and dilbit oil. During the risk characterizations, participants discussed additional, more detailed aspects of the response actions.

4.2.1. Natural Attenuation and Monitoring (NAM)

Natural attenuation relies on natural processes to decrease or "attenuate" concentrations of contaminants (oil) in soil, groundwater, and water. NAM can also be used when the oil is not recoverable, more environmental damage will occur from the response actions, or effective spill response resources are not available. NAM may require extensive monitoring via sampling and other methods, a network of trained observers, detailed sampling protocols, and other unique underwater sampling methods for sinking oils (e.g., dilbit crude oil if it sinks after the majority of light fractions evaporate). Monitoring typically involves collecting soil, groundwater, and water samples to analyze them for the presence of contaminants (oil) and other site characteristics. Some limitations/complications of NAM include: difficulty locating/tracking dilbit if it sinks, potential significant substrate environmental impact if residual dilbit oil sinks, fisheries closings, and public dissatisfaction with the oil spill response. The effectiveness of NAM depends on a multitude of factors such as oil type, ambient weather, and other environmental considerations. Attenuation may be most prudent for Bakken crude oil. Because dilbit is more persistent, it may not be a good candidate for attenuation on land. In responses involving NAM, monitoring would be required (visual monitoring at a minimum) for both oil types; long-term monitoring of spilled oil is most effective and practical on land, compared to a spill in water.

4.2.2. Fire—Let Burn and Controlled Burn (both *In-situ*)

Allowing product to burn *in-situ* is another possible response action. Given the five scenarios, CERA participants considered allowing rail cars to burn themselves out, or control the burn to reduce environmental impacts of the spilled oil. Although (intentional) controlled burning *in situ* was discussed and remained as an option, fire boom would have to be available and the logistics and regulatory approval could preclude its implementation. Air quality issues play a significant role with intentionally burning Bakken oil, e.g., concerns about breathing in the toxins that are in the resultant smoke plume. Example logistical considerations for this response action that must be addressed include:

- How and where to obtain and apply water-cooling streams?
- Are current fireboats sufficient and able to respond?
- Can first responder and public safety air monitoring be deployed?
- Can adequate protection of exposed structures be attained?

Potential limitations of this response action include sufficient access to water supply, frac tanks, high-flow fire pumps and nozzles. The effectiveness of this response action is highly dependent upon the ability to get close enough for effective water cooling, as there is a great potential for "heat-induced tear" in the rail car's shell resulting in a rapid release of vapor and violent fire.

4.2.3. Fire—Extinguishing Agent and Methods

This response action includes the strategies and tactics that use extinguishing agents, including firefighting foam, and the people, equipment and other resources used to extinguish a crude oil fire. Extinguishing agents put a fire out by disrupting one of the four pieces of the fire tetrahedron. Logistical considerations include foam availability and applicability given the incident-specific conditions and consideration of adjacent rail cars or shipboard tanks that must be addressed when utilizing extinguishing agents. Some limitations of using extinguishing agents include:

- The ability to access, in a timely manner,
- Sufficient quantity of foam and water,
- Dry chemical agents,
- Fire boom, and
- Fire boats with the necessary high-volume fire pumps and nozzles.

Additionally, fighting crude oil fires requires highly skilled and specialized personnel. This response action is very effective if the required resources arrive quickly, and the methods are applied properly to the developing situation. This is especially critical before adjacent rail cars are heated to the point of shell plate failure and catch fire.

4.2.4. No Fire—Vapor Suppression

This strategy uses vapor suppression agents, *i.e.*, firefighting foam, to reduce and/or blanket the vapors being released from pooled crude oil in order to reduce the risk of fire and to provide a safe working environment for the first responders and the surrounding public. Examples of logistical considerations include sufficient quantity of foam, regulated products under Subpart J of the NCP (such as herding agents and encapsulators), absorbents, intrinsically-safe vacuum pumps/trucks, personal protective equipment (PPE), and air monitoring equipment. This strategy is not without limitations, because the application of firefighting foam on waters of the US, and where the runoff would enter navigable waters, involves water pollution regulatory issues. Additionally, depending upon the scenario and amount of product spilled, a large coverage area may be required to suppress the vapors. This response action is effective for pooled crude oil in containment (e.g., oil contained by boom, drainage ditch, or small creek).

4.2.5. No Fire—Oil Spread Control (On-Land, On-Water, Underwater)

This response action includes the strategies and tactics that control the spread of oil, and the people, equipment and other resources used to contain the oil. Examples of this strategy include containment and/or deflection boom, sorbents, pneumatic curtains, turbidity curtains, dams/dikes, interceptor trenching, underflow dams, and pre-staged boom. Some limitations of utilizing this action include ensuring a sufficient quantity of boom, sorbents, and other materials given the amount spilled, current weather conditions, and the type of product. Additionally, the speed of deployment is critical to reduce the spread as winds, tides/currents; and ice can hamper response actions. Controlling the spread of submerged oil is a unique and challenging task, and may require non-typical oil spill response techniques. These response actions can be very effective if deployed correctly and in a timely manner. For underwater oil spread control, turbidity, silt, and pneumatic curtain effectiveness may be impacted by surface and subsurface currents and tidal exchanges, and are sometimes difficult to position and hold in place with changing environmental conditions. Additionally, extreme tidal ranges, which may expose mud flats at low tide will decrease the efficiency of this response action due to the difficulty of deploying and maintaining floating boom under these conditions.

4.2.6. No Fire—On-Water Recovery and Underwater Recovery

This response action is used to recover spilled crude oil from the water's surface or subsurface, for the purpose of preventing oiling and minimizing damage to sensitive shoreline resources and habitats. Example resources used to enact this strategy include skirted booms, self-propelled skimmers, stationary skimmers, and advancing skimmers (brush, drum, weir, and Dynamic Inclined Plane/DIP), dredges (hydraulic, clam shell), trawls, nets, and vacuum systems. Both on-water and underwater recovery is limited by a number of factors including the type of skimmer, which must be selected for the type of oil and the weather conditions, and the ability to access the oil. Recovery amount is dependent upon many factors, one of which is encounter rate, or the area of oil that an individual skimmer can encounter over a period of operational time e.g., in 12 h per day. Brush skimmers have been shown to be more effective and efficient for heavy oils like dilbit; assuming the oil remains floating. DIP, weir, filter belt, disk, and drum skimmers have been shown to be more effective for recovering light oils like Bakken. Size and configuration of these skimmers must be commensurate with the weather and sea conditions to achieve maximum effectiveness.

Additionally, collection booming, nets, trawls, pumps, dredges, divers, vacuum systems, airlifts, and bottom trawls can be used to recover heavy oils, *i.e.*, those with a specific gravity equal to or heavier than water from being produced that way, e.g., some #6 oils, or attains a specific gravity equal to or heavier than water through physical or chemical changes (weathering). After the lighter components of dilbit evaporate, the heavier and more viscous components remain on the water. Under certain conditions, the remaining oil can pick up sediment from the water column resulting in increased specific gravity. An increase in specific gravity can result in oil that becomes neutrally buoyant or heavier than water, causing it to submerge below the water's surface. Oil in this state is very difficult to locate and recover. Diver effectiveness is impaired by low visibility, and differentiating oil from mud. Recovery of submerged oil in rivers and estuarine areas with heavy sediment load and currents is especially challenging. The use of remote sensing and GPS integrated systems can increase the effectiveness of underwater recovery.

4.2.7. No Fire—Resource Protection (On-Water and On-Land)

This response action involves protecting sensitive areas by deploying protection strategies using boom, which are physical barriers used on land or water (floating), made of plastic, metal, or other materials, which slow the spread of oil and keep it contained. Boom can also be utilized to deflect oil away from sensitive areas, to include water intakes, historic sites, and critical fishery areas. Types of floating, skirted boom (cylindrical float at the top and is weighted at the bottom so that it has a "skirt" of varying dimensions under the water) considered include: 12" boom for protection/deflection due to shallow water in rivers and creeks, and ease of use; 18" boom used for deeper water areas like bays and inlets; and larger (24"+) boom used for coastal and offshore areas. A turbidity/silt curtain can be used to limit submerged oil movement. Other protection methods include pneumatic curtains, dams/dikes, interceptor trenching, underflow dams, and shore-seal boom. Although there are many types of equipment and tactics to protect sensitive areas, shoreline type, oil type and volume, topography, porosity, and shape will limit the effectiveness of protection strategies. Stakeholders should recognize that protection of 100% of shorelines and sensitive areas is impractical, if not impossible. The tidal range and shallowness of some creeks and tributaries expose the mud flats at low tide; therefore restricting protective boom deployment to higher-tide hours only, and may impact placement of protection boom in general. Additionally, current and tide necessitates that boom be tended at every tide cycle. Protection strategies for floating oils have been demonstrated to be effective, when anchored properly and tended round-the-clock. It is important to note that mechanical protection of large areas, e.g., around or in front of islands or in across the mouth of a bay, is much more difficult that lay people imagine. Effective protection strategies for non-floating oils are even more difficult to implement.

4.2.8. No Fire—Shoreline Clean-up

The use of this strategy involves the removal of oil from the shoreline for long-term disposal elsewhere to prevent further or introduction of contamination to sensitive areas and habitat. Examples of this response action include mechanical recovery systems (vacuum trucks, storage tanks, sorbent, hand tools, laborers), and NCP Subpart J surface washing agents. To reduce the amount of recovered oily waste for disposal, pre-spill impact debris removal is advised, which could be quite extensive depending upon the location. The specific shoreline cleanup method selected will be based on shoreline type and oil type; shoreline access, and consideration that habitat may be affected detrimentally by the cleanup activity (foot traffic or machinery) itself. The effectiveness of shoreline cleanup depends on many factors: oil type (heavy *vs.* light), type of shoreline and the amount of debris present, and the fact that tidal ranges and cycles can significantly impact responder work schedules (e.g., daylight hours and total time shoreline is exposed at low tide). A shoreline cleanup response requires close coordination between the personnel conducting response operations and Shoreline Cleanup (or Countermeasure) Assessment Technique (SCAT) teams to determine extent of shoreline contamination, cleanup priorities, and acceptable methods for removing the oil. Depending on the magnitude and location of the spill, shoreline cleanup can be the most logistically demanding portion of an oil spill response. Shoreline cleanup can have a high degree of collateral damage.

4.2.9. No Fire—Oil Detection/Mapping (Physical-Contact Methods)

Oil detection and mapping includes the strategies, methods, and resources used to detect oil by physically sampling habitats to track oil movement, location, physical properties, and extent of contamination. Examples of this response action include collecting water, soil and air samples, and establishing monitoring stations and other oil detection sensors. Examples of sub-surface measuring techniques include using trawls and underwater sentinels, crab pots, snare samplers, Vessel Submerged Oil Recovery Systems (VSORS), and remotely-operated vehicles (ROVs). These methods can result in physical trauma of the habitat, and associated organisms. There are many limitations to this response action, especially if the oil is submerged, such as the need to rapidly develop a complete and defensible sampling protocol and procedures to adequately check potentially-impacted areas, given that tides, winds and currents spread oil quickly. Determining sampling locations may be difficult for submerged oil, especially if the oil is mobile. Effectiveness of these response actions is limited by the lack of a full suite of effective technologies to detect oil and map its extent of contamination in all subsurface environments (water column or benthos). Currently, some subsurface mapping methods exist, but this remains an active research and development area.

4.2.10. No Fire—Oil Detection/Mapping (Remotely-Observed Methods)

This response action involves the use of remotely-operated sensors and human vision to detect and monitor the movement of oil in the environment from a distance, in this case, on the water surface or subsurface. Examples of remote sensing include visual observation via overflight and technological sensors including laser sensors, infrared, and photobathymetric sensors. On-water remote sensors include sonar scans (e.g., side-scan, multi-beam, *etc.*) laser fluoro-sensors, and underwater visual detection by divers or remotely operated vehicles and autonomous underwater vehicles. Limitations of these techniques include weather interference, adequate detail to inform decision making, e.g., thickness of oil slicks and mistaking oil for other substances), availability of equipment and operators, data interpretation, and the development of comprehensive sampling protocols and procedures Effectiveness of these response actions is limited by the lack of a full suite of effective technologies in all ambient conditions (ice, poor visibility in air or in the water) to detect oil and map its extent of contamination in all surface and subsurface environments (water column or benthos). Some methods to detect oil remotely this do exist, but this also remains active research and development area. There is usually little to no collateral damage with these methods but their effectiveness can also be limited by incident-specific conditions.

4.3. Conceptual Model

The conceptual model considers the risk of potential exposure of a resource to the spilled oil, recognizing that a CERA is qualitative, not quantitative, in nature. In the US, quantitative assessments of the ecological severity and extent of adverse impacts on natural resources and the services they provide, as a result of the spilled oil, are conducted by the Natural Resources Trustees. That quantitative assessment is coordinated with, but separate from, response actions during an incident. The conceptual model developed for this CERA depicts the connections between the resources of concern (human health, socio-economic, and ecological) and their potential to be exposed to hazards (exposure pathway) for five scenarios.

The definitions of hazards used in completing the conceptual models for each of the scenarios are:

1. Air pollution—vapors, direct effects from respiratory issues for air breathers. Therefore, air pollution is not a stressor for mid-water, benthos.
2. Aqueous exposure—direct effects from aquatic respiration and dermal exposure to oil and oil components dissolved within the water column; may be short-lived exposure with the potential for high consequence for impacted species. Excludes submerged oil globules.
3. Physical trauma (mechanical impact from equipment, aircraft, people, boats, *etc.*)—direct effects from physical impact on individual species, including disturbance.
4. Oiling/smothering—direct effects from dermal contact with oil; skin (hypothermia), mucosal membranes (eyes, nares, *etc.*); indirect effects or secondary impacts could include ingestion (preening). This may include contact with submerged oil globules or mats.
5. Thermal (heat exposure from fire)—direct effects from oil burning; impacts from exposure to a fire/burn (not dermal exposure to the oil).
6. Waste—direct effects prior to being removed (pre-cleaning) from the system. Excludes equipment intended for re-use, e.g., non-sorbent boom.
7. Ingestion (food web, *etc.*)—resources indirectly exposed to oil or its constituents via ingestion of oil or contaminated/affected prey.
8. Advisory/Closure—prohibited action of use (e.g., commercial or recreational fishery, water intake); protection from possible exposure.

Conceptual models were developed for human health and safety, plus five ecological models were developed for each scenario (all workgroups completed these separately). They are included in the project report [38]. The numbers in the matrix cells (1–8) indicate the path by which a hazard can affect a resource. The completed model for each scenario presents the participants' decision making and reasoning for each scenario workgroup about the concern for each resource. *NA* represents the absence of a connection between a potential hazard and the resource of concern as determined by the workgroup participants for the individual scenario.

Recognizing the distinctive oil behaviors of Bakken crude and dilbit oils, the conceptual models developed for these oils reflect the two phases of potential exposure pathways:

- B1: Bakken on the water/soil surface (initially, as it begins to weather)
- B2: Bakken within water column due to natural dispersion (later)
- D1: Dilbit before weathering (initially, from loss of lighter fractions through evaporation)
- D2: Dilbit after weathering (later)

4.4. Risk Ranking Matrix

The CERA uses a risk ranking matrix to assign levels of concern about the potential severity and duration of impacts caused by the spilled oil, if left to attenuate naturally, or as addressed by the individual response actions listed in Section 4.2. After reviewing the risk matrix from previous CERAs, a modification was adopted for evaluating the relative levels of concern of the impacts of Bakken or dilbit oils in the five scenarios. The Y-axis of the risk ranking matrix shown in Figure 6, was used to describe the collectively perceived ecological severity, rather than the percent loss/reduction of the specific natural resource based on a measured endpoint that is not site-specific/species-specific/community-specific to the instant scenarios and were not always applicable to the selected scenarios. The X-axis in Figure 6 describes Recovery Time over arbitrary, but participant-consensus based, periods of time. Use of the matrix is qualitatively dependent on the individual participant's experience and perception with respect to the complexity of the habitats and species present. Notwithstanding, the final ranking of risk is achieved through consensus of the participants thereby arriving at an estimation of the risk that is acceptable to multiple parties.

		RECOVERY TIME			
		RAPID	MODERATE	MODERATE	SLOW
		< 1 year (4)	1 to 4 years (3)	5 to 10 years (2)	> 10 YEARS (1)
	Discountable (D)	4D	3D	2D	1D
ECOLOGICAL SEVERITY	Impaired (C)	4C	3C	2C	1C
	Significant (B)	4B	3B	2B	1B
	Dysfunctional (A)	4A	3A	2A	1A

Figure 6. Levels of concerns risk matrix used in the Bakken and dilbit oil CERA. [■] Limited Level of Concern; [■] Moderate Level of Concern; [■] High Level of Concern.

Participants evaluated potential risks of the oil and response actions and assigned levels of concern using the best available information from literature, past experience with oil spills in an area, e.g., the 2004 oil spill from the M/V Athos-1 in the Delaware River, and their knowledge of resources in the area, rather than additional data collection or field studies. As can be seen in the risk matrix, the groups used alphanumeric scores to scale the anticipated impact severity and recovery time. After developing the scaling, color coding was used to indicate the summary levels of concern. The resulting risk scores represent a participant consensus that severity and duration of consequences were likely to occur in the given scenario.

Risk is defined as the probability of an impact occurring. Participants qualitatively considered if there was a high, medium, or low probability of the impact occurring, and then determined the severity and the duration of the impact. During an incident, responders tend to consider and decide about impacts that are relatively short-term and based upon what was learned from previous response events, e.g., it is better to protect marshes to avoid having them oiled because of the longer-term severity and duration of oil impacts in those environments. The response community recognizes that protecting marshes from oiling is a best response management practice; this knowledge proactively and affirmatively guides many preparedness and response decisions. For any T/E species, any harm qualifies as a take under the Endangered Species Act, Section 7, and Incidental Take of Endangered and Threatened Species in U.S. Lands or Waters. Being listed on the ESA makes it illegal to take. Take is defined as harass, harm, pursue, hunt, shoot, wound, kill, trap, capture, collect, or attempt to do these things (50 CFR § 3(19), 2009) any of these protected species, whether endangered or threatened or adversely modify or destroy designated critical habitat under Section 9—Prohibited Acts. These prohibitions under Section 9 are not automatic for threatened species; the USFWS and NMFS must conduct a Section 4 process to address threatened species. and is significant. As generally applied in this ERA, non T/E plants would recover in four to five years and non T/E fish would recover in one to two years. The ranking of red does not mean to stop response actions, but rather to review and assure that response actions would not adversely affect the species of concern, that the risk is recognized, and deemed appropriate to decision makers, including resource managers.

The definitions of ecological severity used by participants in assessing risk in this CERA are:

- Discountable: Impacts are considered negligible, trivial, or a minor inconvenience.
- Impaired: Short-lived modestly adverse impacts that alter habitats or life cycles.
- Significant: Sustained and substantive adverse impacts that potentially lethal or highly damaging to a natural resource(s).
- Dysfunctional: Long term damage that prohibits a natural resource from living, reproducing, or providing an ecological service(s).

The ranking measures were also appropriate for characterizing human health risks. The impact from a drinking water ban would likely be considered *Dysfunctional* in severity; inhalation and dermal impacts might be ranked as *Significant*.

Duration of impact begins from the time of the oil discharge. Severity takes into account the significance of individual organisms relative to the scale of population. For example, if an organism has recovered 70% in about one year, but would take 10 years for 100% recovery, then the risk could be ranked as significant, in the one to four-year duration. Local populations that could be killed by a spill would receive a dysfunctional score. Freshwater mussels, for example, if wiped out by oiling in a creek, will not recover for 50–100 years. This would equate to a risk of a dysfunctional impact. Generally, participants considered populations of organisms at the local scale, and assumed no impacts on a regional or national scale for that species.

4.5. Risk Characterization

The conceptual models were used first to clarify the pathways of exposure and the types of hazards between the spilled oil and seasonally-present ecological resources. The risk matrix was first completed by each workgroup to characterize the risks to resources of concern from the oil only, *i.e.*, no response action, except for natural attenuation and monitoring (NAM). Next, participants compared the potential risks of each category of response actions to the risks associated with the spilled oil left in place to attenuate, plus monitoring via sampling. Relative risks were compared in this way:

- If using a response action is likely to *improve* the outcome, the score is a *lower* alphanumerical value than the spilled oil (NAM).
- If using a response action is likely to *worsen* the outcome, the score is a *higher* alphanumerical value than the spilled oil (NAM).

Selecting appropriate response strategies for each of these oils takes into account their behavior in the environment before and after weathering. For this reason, Figures 7 and 8 display the summarized risks for the oils and strategies for the initial spill of oil before it has a chance to weather and Figures 9 and 10 display the summarized risks for the oils and strategies for the oil after it has weathered. For convenience, a row that highlights the transportation setting (*i.e.*, urban, creek, river or bay) of each of the scenarios has been added to the summary tables.

Habitat	Artificial Shorelines			Natural Terrestrial Shorelines			Intertidal Shoreline (Exposed & Sheltered) / Surface Water (0 - 1			Mid-water (0 to 2 meters)			Benthic (bottom, >2 meters)		
Sub habitats	Bulkheads, Riprap, Manmade Structures, Pavement			Vegetated,Grass, Sand, Gravel			Marsh, Swamp, Tidal flats, Sand Beaches, Cobble/Boulder Beach			Water Column			Bottom of the Water Column / Seabed		
SCENARIO	1	2	3	1	2	3	1	2	3	1	2	3	1	2	3
Transportation Setting	Urban	River	Bay	Urban	River	Bay	Urban	River	Bay	Urban	River	Bay	Urban	River	Bay
Response Actions															
Oil Only: No response action except monitoring (air: flammability, benzene, etc.; soil; water; and stranded onshore oil)															
Fire															
Let burn and controlled burn (both in-situ)															
Extinguishing agents and methods															
No Fire															
Vapor suppression															
Oil spread control															
On-water oil recovery															
Resource protection															
Shoreline clean up															
Oil Detection/mapping - remotely observed methods															

Figure 7. Summary Risk Characterization for the Bakken oil scenarios (initial release). █ Limited Level of Concern; █ Moderate Level of Concern; █ High Level of Concern; █ Not Applicable.

Habitat	Artificial Shorelines		Natural Terrestrial Shorelines		Intertidal Shoreline (Exposed & Sheltered) / Surface Water		Mid-water (0 to 2 meters)		Benthic (bottom, >2 meters)	
Sub habitats	Bulkheads, Riprap, Manmade Structures, Pavement		Vegetated, Grass, Sand, Gravel		Marsh, Swamp, Tidal Flats, Sand Beaches,		Water Column		Bottom of the Water Column / Seabed	
SCENARIO	4	5	4	5	4	5	4	5	4	5
Transportation Setting	Creek	River	Creek	River	Creek	River	Creek	River	Creek	River
Response Actions										
Oil Only: No response action except monitoring (air: flammability, benzene, etc.; soil; water; and stranded onshore oil)										
Fire										
Let burn and controlled burn (both in-situ)										
Extinguishing agents and methods										
No Fire										
Vapor suppression										
Oil spread control										
On-water oil recovery										
Resource protection										
Shoreline clean up										
Oil Detection/mapping - physical detection/remotely observed methods										

Figure 8. Summary Risk Characterization for the dilbit oil scenarios (initial release). █ Limited Level of Concern; █ Moderate Level of Concern; █ High Level of Concern; █ Not Applicable.

Habitat	Artificial Shorelines			Natural Terrestrial Shorelines			Intertidal Shoreline (Exposed & Sheltered) / Surface Water (0 - 1			Mid-water (0 to 2 meters)			Benthic (bottom, >2 meters)		
Sub habitats	Bulkheads, Riprap, Manmade Structures, Pavement			Vegetated, Grass, Sand, Gravel			Marsh, Swamp, Tidal flats, Sand Beaches, Cobble/Boulder Beach			Water Column			Bottom of the Water Column / Seabed		
SCENARIO	1	2	3	1	2	3	1	2	3	1	2	3	1	2	3
Transportation Setting	Urban	River	Bay	Urban	River	Bay	Urban	River	Bay	Urban	River	Bay	Urban	River	Bay
Response Actions															
Oil Only: No response action except monitoring (air: flammability, benzene, etc.; soil; water; and stranded onshore oil)															
Fire															
Let burn and controlled burn (both in-situ)															
Extinguishing agents and methods															
No Fire															
Vapor suppression															
Oil spread control															
On-water oil recovery															
Resource protection															
Shoreline clean up															
Oil Detection/mapping - remotely observed methods															

Figure 9. Summary Risk Characterization for the Bakken oil scenarios (weathered). [ ] Limited Level of Concern; [ ] Moderate Level of Concern; [ ] High Level of Concern; [ ] Not Applicable.

Habitat	Artificial Shorelines		Natural Terrestrial Shorelines		Intertidal Shoreline (Exposed & Sheltered) / Surface Water		Mid-water (0 to 2 meters)		Benthic (bottom, >2 meters)	
Sub habitats	Bulkheads, Riprap, Manmade Structures, Pavement		Vegetated, Grass, Sand, Gravel		Marsh, Swamp, Tidal Flats, Sand Beaches, Cobble/Boulder Beach		Water Column		Bottom of the Water Column / Seabed	
SCENARIO	4	5	4	5	4	5	4	5	4	5
Transportation Setting	Creek	River	Creek	River	Creek	River	Creek	River	Creek	River
Response Actions										
Oil Only: No response action except monitoring (air: flammability, benzene, etc.; soil; water; and stranded onshore oil)										
Fire										
Let burn and controlled burn (both in-situ)										
Extinguishing agents and methods										
No Fire										
Vapor suppression										
Oil spread control										
On-water oil recovery										
Resource protection										
Shoreline clean up										
Oil Detection/mapping - physical detection/remotely observed methods										

Figure 10. Summary Risk Characterization for the dilbit oil scenarios (weathered). [ ] Limited Level of Concern; [ ] Moderate Level of Concern; [ ] High Level of Concern; [ ] Not Applicable; [ ] Unable to Determine Due to Insufficient Information.

Bakken and dilbit oils present flammability hazards during the early stages of a spill, such as the 2013 derailment in Lac-Megantic, Canada that resulted in Bakken oil fire in the center of town and loss of life. The risks to human health and safety and social-economic resources from a fire, as well as firefighting foam, as highlighted in Scenario 1, were scored as a moderate level of concern (yellow) assuming that safety measures were successful, e.g., PPE for workers and other protective measures for the public, such as safe distance from fire and shelter in place away from the smoke plume. For human health and safety, the levels of concern from the oil only and for other response actions were scored as low (green).

The red ranking (high level of concern) represents a need for follow-up action by the USCG and other agencies in the area to more fully address whether certain response actions will be allowed and if so, under what circumstances, e.g., in certain seasons or under specific conditions. The red ranking, highest relative risk, is not intended to prevent or stop response actions, but rather to prompt further review and assure that response actions would not adversely affect the resources of concern. In this CERA, participants generally assigned higher levels of concern about the ecological risks associated with a spill of dilbit oil compared to a spill of Bakken crude oil.

The gray cells scored "Not Applicable" refer to the absence of either the resource of concern in a scenario or the lack of pathway for exposure to the hazard presented by the oil or type of response action. For example, in Scenario 3, which involves a spill of Bakken oil in the upper Delaware Bay in winter, during the first 4–6 h, the risk to shorelines and benthos were scored as Not Applicable because the oil would not reach those environments in that time frame.

4.5.1. Bakken Oil Spill Risks

Scenario 1 occurs at a rail crossing within Philadelphia's urban setting; Scenario 2 involves a barge in the middle of the Delaware River near Pea Patch Island, an ecologically and culturally-sensitive resource; and Scenario 3 involves a tanker in the open water of upper Delaware Bay.

In general, the workshop participants from Scenarios 1, 2, and 3 found that the "No Response other than monitoring" option was considered of limited or moderate level of concern when deemed appropriate for the scenario conditions, both at the 4 to 6 h response frame or at four to seven days post discharge. In most cases, the participants found that there was very little change in concern levels when considering the various response action *versus* the NAM action. Although some of the levels of concern did increase from low to moderate (green to yellow), this increase in concern did not necessitate a change in response options. The highest level of concern in the urban, freshwater scenario (1), was with the use of extinguishing or vapor suppression agents in the intertidal zone for both the initial response and over the four to seven day response times.

Overall, the highest level of ecological concern (red) in these scenarios occurs from the risk of the use of firefighting foam or vapor suppression agents to threatened/endangered species, which might be present in Scenario 1 intertidal shoreline. The use of foam in this area increased the risk (yellow) over the presence of oil only (green). The runoff from firefighting foam, if applied, could present both human health and ecological risks. Many earlier formulations of fire suppression foam contain perfluorochemicals (*PFCs*) that were used to improve smothering capability; however these formulation are being phased out in the US [48]. Specific actions that decreased or increased the risks to resources of concern over the oil alone, *i.e.*, NAM, are discussed below.

In the initial 4–6 h after a Bakken spill occurs in these scenarios, the response actions that positively change, *i.e.*, decrease the risk (change a yellow to a green score) compared to the NAM are: Scenario 1—oil spread control, on-water recovery, resource protection, shoreline cleanup, and remotely observed oil detection/mapping in the mid-water habitat (Scenario 1). In the initial 4–6 h after a Bakken spill occurs in these scenarios, the response actions that negatively change, *i.e.*, increase the risk (change a green to a yellow score) compared to NAM are: Scenario 1—oil spread control, on-water recovery, resource protection, shoreline cleanup, and remotely observed oil detection/mapping in natural terrestrial shorelines.

In Scenarios 2 and 3, implementing response actions does not noticeably decrease ecological risks compared to the risk of NAM. After the oil has been in the environment for four to seven days, much of the Bakken oil would have weathered, leaving a light residual oil staining on shorelines, and the intertidal portion of the shoreline, which could be below the water surface (surface water or mid-water column habitats) during high tide. After the oil has been in the environment for four to seven days, the response actions that positively change, *i.e.*, decrease the risk (change a yellow to a green score) compared to NAM is: Scenario 2—implementing on-water oil recovery near intertidal shorelines. After the oil has been in the environment for four to seven days, the response actions that negatively change, i.e., increase the risk (change a green to a yellow score) compared to oil only are: shoreline cleanup in artificial shorelines and natural terrestrial shorelines in Scenarios 1 and 2.

4.5.2. Dilbit Oil Spill Risks

Scenario 4 occurs at a rail crossing over Mantua Creek in New Jersey; Scenario 5 involves a barge in the middle of the Delaware River near the Marcus Hook Anchorage. Generally, the ecological risks associated with spilled dilbit oil and the anticipated response actions for the two scenarios are either moderate or high level of concern, especially if T/E species are present in contaminated areas.

As with Bakken oil, the highest level of concern (red) in these scenarios occurs with threatened/endangered species, which may be present and could be impacted by the oil and/or response actions. In these scenarios, T/E species (e.g., Atlantic sturgeon to sea turtles to bald eagles) might be present and at risk in all environments except artificial shorelines.

In general, the workshop participants from Scenarios 4 and 5 found that the NAM option provided a moderate to high level of concern, both at the 4 to 6 h response timeframe to four to seven days post-discharge. In most cases, the participants found that there was very little change in concern levels when considering the various response action *versus* the NAM action. Some of the levels of concern did increase for response actions over the 4 to 6 h timeframe to the four to seven day period from low to moderate (green to yellow). The high level of concerns (red) were scored for scenario 5 regarding the use of most response options deemed applicable for use in the intertidal, mid-water, and benthic zones for both the initial response and over the four to seveb day response times. Similar concerns were also expressed for Scenario 4 when considering the risk of shoreline clean up and oil detection/mapping methods for the intertidal and midwater zones. Specific actions that decreased or increased the risks to resources of concern over the oil alone, *i.e.*, NAM, are discussed below.

In the initial 4–6 h after a dilbit spill occurs in these scenarios, the response actions that positively change, *i.e.*, decrease the risk (change a yellow to a green score, or a red to a yellow score) compared to NAM are: Scenario 5—oil spread control and on-water recovery in artificial shorelines and at the water's edge (0 meters of the mid-water habitat); and resource protection in intertidal shorelines. In the initial 4–6 h after a dilbit spill occurs in these scenarios, NAM actions negatively change, *i.e.*, increase the risk (change a green to a yellow score, or a yellow to a red score) compared to the presence of the NAM action only.

It is important to note that in some habitats, oil detection and mapping methods (if physically disturbing) were scored as a higher risk (red) than oil spread control, on-water recovery and resource protection to natural vegetated shorelines, intertidal shorelines, and mid-water habitats (yellow or green). Remote sensing methods would present a low (green) ecological risk. Participants recognized that available methods to detect and recover submerged dilbit oil are lacking in effectiveness.

Workgroup participants noted that on-water oil recovery in the earliest stage after release is the most important response action to prevent the oil from spreading out and expanding the extent of contamination, e.g., into the Delaware River, and contaminating larger shoreline and benthic habitats and organisms. This oil begins as a crude oil which flows but weathers to a heavy oil product that resembles #6 oil that will be extremely tacky and adhesive. The oil is expected to behave differently as it weathers; it could behave as one type of weathered product and a different type a few hours later. After the light ends have volatized, the residual product will adhere to whatever it contacts. From an environmental standpoint, this is a significant challenge for onshore cleanup and wildlife rehabilitators to avoid if at all possible. It will take time to experiment (trial/error) with emerging ideas and techniques and develop new response/restoration best management tactics to manage dilbit releases. This was the situation during the 1989 M/V Presidente Rivera oil spill on the Delaware River, during which the set of response tactics varied in the same day. In the morning when the oil temperature was below the pour point and did not flow on beach sand, it could be recovered with a pitchfork, for example. However, in the afternoon when the sunlight warmed the oil stranded on the sand and it flowed, it could no longer be picked up, and sorbent booms were needed to contain it.

After four to seven days, much of the dilbit oil would have weathered to the point that the light fractions had evaporated, leaving the heavier, persistent bitumen in the environment, available to pick up sediments in the water column and sink to the benthos, or on shore. After the oil has been in the environment for four to seven days, the response action that positively changes, *i.e.*, decreases the risk (change a yellow to a green score) compared to the remaining oil (NAM only) are:

- Scenario 4—controlled *in-situ* burning in natural terrestrial shorelines, intertidal shorelines, and the water's edge of the mid-water column.
- Scenario 4—oil spread control, on-water recovery, resource protection, shoreline cleanup in natural terrestrial shorelines
- Scenario 5—resource protection in intertidal shorelines.
- Scenario 5—oil spread control, on-water recovery the water's edge of the mid-water column.

After the oil has been in the environment for 4–7 days, the response actions that negatively change, i.e., increase the risk (change a green to a yellow score) compared to NAM are:

- Scenario 5—shoreline cleanup in artificial shorelines (increased risk to reptiles, amphibians, and macro-invertebrates).

The focus of this CERA was to evaluate different response actions to spills involving Bakken and dilbit, which have become part of the USCG's evolving responsibilities in the recent domestic energy renaissance, particularly in the Delaware Bay Watershed. Drawing on a cross-section of the local oil spill response community the ecological impacts of spilled oil and spilled oil plus a response action were evaluated to predict the severity and duration of adverse impacts to natural resources of concern. Once oil spilled into the environment one objective of the emergency response community is to react in such a manner as to minimize or prevent additional harm to the environment. CERA promoted agreement among and between the risk management team, stakeholders, and interested parties towards the common goal of successfully planning for and responding to oil spills in such manner as to transparently consider the conservation of natural resources for the benefit of human uses and wildlife habitat.

This CERA is consistent with this concept of conservation of natural resources while allowing multiple uses. Further work is needed to address domestic energy transportation needs that are compatible with and provide protection of natural resources and the human and ecological services they provide.

Acknowledgments: The authors gratefully acknowledge the willingness and substantial contributions of all those who participated in the CERA, and especially those who provided introductory presentations and led each of the scenario workgroups. This work was funded by USCG Headquarters Contract Number (HSCG23-11-A-MPP439) for U.S. Coast Guard Headquarters (G-RPP), STOP 7516, Office of Marine Environmental Response Policy, 2703 Martin Luther King Jr. Ave. SE, Washington, DC 20593-7516. No funding was obtained to prepare this article.

Author Contributions: The primary author, Ann Hayward Walker, was the consultant representative responsible for the application and facilitation of the CERA process for this oil spill response issue. Clay Stern was the lead representatives for trust resources of the US Fish and Wildlife Service; he also updated the risk ranking matrix and drafted the list of ecological resources of concern. Debra Scholz developed the two-phased weathering approach for the conceptual model and facilitated the update to the draft hazards list and resources of concern. Eric Nielsen provided the USCG background, perspective, and context for this CERA; he also contributed to the descriptions of response actions. Frank Csulak provides scientific support to the USCG FOSC, and was the lead representative for NOAA's trust resources. Rich Gaudiosi developed the description of response actions for this CERA. The views expressed herein are those of the authors and are not to be construed as official or reflecting the views of the U.S. Government or of the respective agencies.

Conflicts of Interest: The authors declare no conflict of interest. The funding sponsors at USCG Headquarters had no role in the design of the study, in the writing of the manuscript, or in the decision to publish the results. A USGC Coast Guard headquarters representative was one of the 88 participants in the CERA.

Abbreviations

The following abbreviations are used in this manuscript:

BPD	Barrels per day
bbls	Barrels
CERA	Consensus Ecological Risk Assessment
DE	Delaware
DIP	Dynamic Inclined Plane
EPA	US Environmental Protection Agency
ERA	Ecological Risk Assessment
Gals	Gallons
NAM	Natural Attenuation and Monitoring
NCP	US National Oil and Hazardous Substances Pollution Contingency Plan
NGO	Non-governmental organization
NJ	New Jersey
NOAA	National Oceanic and Atmospheric Administration
PA	Pennsylvania
PPE	Personal Protective Equipment
T/E	Threatened and endangered species
EIA	US Energy Information Administration
USCG	US Coast Guard
USFWS	US Fish and Wildlife Service

References

1. National Academies of Science, Engineering, and Medicine. *Spills of Diluted Bitumen from Pipelines: A Comparative Study of Environmental Fate, Effects, and Response*; The National Academies Press: Washington, DC, USA, 2016.
2. American Petroleum Institute. Oil Sands. Available online: http://www.api.org/~/media/Files/Oil-and-Natural-Gas/Oil_Shale/Oil_Shale_Factsheet_1.pdf (accessed on 14 January 2016).
3. U.S. Energy Information Administration. *U.S. Crude Oil and Natural Gas Proved Reserves, 2014*; U.S. Department of Energy: Washington, DC, USA, 2015; pp. 1–20.
4. Frailey, F.W. All Oiled Up. How Railroads Got Back in the Oil Business and Why They Aren't Going—Ever. *Trains* **2014**, *74*, 28–37.
5. Fingas, M. Diluted Bitumen (Dilbit): A Future High Risk Spilled Material. Available online: http://interspill.org/previous-events/2015/WhitePapers/Interspill2015ConferenceProceedings/25%20MARCH%202015/Future%20Risk%20-%20HNS%20-%20Bitumen/Diluted%20-Bitumen%20-Dilbit-A-Future-High-Risk-Spilled-Material.pdf (accessed on 28 February 2016).
6. Mitchell, J.H.; Child, N.J. *Bakken Crude Oil Spills-Response Options and Environmental Impacts*; MassDEP: Canton, MA, USA, 2015; pp. 1–100.
7. United States Environmental Protection Agency. ExxonMobil Mayflower Clean Water Settlement. Available online: http://www.epa.gov/enforcement/exxonmobil-mayflower-clean-water-settlement (accessed on 14 January 2016).
8. Brescia, N.; Dodson, A. ExxonMobil Mayflower Oil Spill. Available online: http://rrt6.org/Uploads/Files/05-29%201430%20ExxonMobil%20Pipeline%20Mayflower%20Oil%20Spill.pdf (accessed on 14 January 2016).
9. Pipeline and Hazardous Materials Safety Administration. ExxonMobil Pipeline Incident-Mayflower, Ark. Available online: http://www.phmsa.dot.gov/portal/site/PHMSA/menuitem.6f23687cf7b00b0f22e4c6962d9c8789/?vgnextoid=1a9ab5676d5cd310VgnVCM100000d2c97898RCRD&vgnextchannel=d248724dd7d6c010VgnVCM10000080e8a8c0RCRD&vgnextfmt=print (accessed on 14 January 2016).
10. Wikipedia. 2013 Mayflower Oil Spill. Available online: https://en.wikipedia.org/wiki/2013_Mayflower_oil_spill (accessed on 14 January 2016).
11. NOAA's Response and Restoration Blog. Five Years After Deepwater Horizon, How Is NOAA Preparing for Future Oil Spills? Available online: https://usresponserestoration.wordpress.com/tag/tar-sands/ (accessed on 14 January 2016).

12. Doelling, P.; Davis, A.; Jellison, L.T.K.; Miles, S. Bakken Crude Oil Spill Barge E2MS 303 Lower Mississippi River February 2014. Available online: http://www.nrt.org/production/NRT/RRT3.nsf/Resources/May2014_pdf/$File/Bakken_Crude_Spill_E2MS303_Revised.pdf (accessed on 14 January 2016).

13. NOAA. Barge E2MS 303: IncidentNews. Available online: http://incidentnews.noaa.gov/incident/8729 (accessed on 14 January 2016).

14. Karlamanga, S. Train in Alabama Oil Spill Was Carrying 2.7 million Gallons of Crude. Available online: http://articles.latimes.com/2013/nov/09/nation/la-na-nn-train-crash-alabama-oil-20131109 (accessed on 13 January 2016).

15. Environment Canada; Fisheries and Oceans Canada; Natural Resources Canada. *Properties, Composition and Marine Spill Behaviour, Fate and Transport of Two Diluted Bitumen Products from the Canadian Oil Sands*; Government of Canada: Quebec, Canada, 2013; pp. 5–80.

16. Polaris Applied Sciences, Inc. A Comparison of the Properties of Diluted Bitumen Crudes with Other Oils. Available online: http://crrc.unh.edu/sites/crrc.unh.edu/files/ comparison_bitumen_other _oils_polaris_2014.pdf (accessed on 14 January 2016).

17. Fitzpatrick, M.; Tebeau, P.; Hansen, K.A. *Development of Bottom Oil Recovery Systems-Final Project Report*; USCG: Acquisition Directorate Research & Development Center: New London, CT, USA, 2013; pp. 1–35.

18. Witt O'Brien's; Polaris Applied Sciences; Western Canada Marine Response Corporation. *A Study of Fate and Behavior of Diluted Bitumen Oils on Marine Waters*; Dilbit Experiments: Gainford, AB, Canada, 2013; pp. 1–66.

19. Auers, J.R.; Couture, R.M.; Sutton, D.L. *The North Dakota Petroleum Council Study on Bakken Crude Properties*; Prepared for the North Dakota Petroleum Council. Dallas, TX, USA, 2014; pp. 1–78. Available online: http://www.ndoil.org/image/cache/Bakken_Quality_Report.pdf (accessed on 28 February 2016).

20. Transportation Safety Board of Canada. Laboratory Report LP148/2013. Available online: http://www.tsb.gc.ca/eng/enquetes-investigations/rail/2013/R13D0054/lab/20140306/LP1482013.asp (accessed on 1 January 2016).

21. U.S. Environmental Protection Agency, Emergency Response Team. Bakken Shale Crude Oil Spill Evaluation Pilot Study. Available online: http://rrt5.org/Portals/0/docs/BakkenCrudePilot Study-PrelimRpt_Final_4-1-15.pdf (accessed on 14 January 2016).

22. New York State Departments of Environmental Conservation, Health, & Transportation; Division of Homeland Security and Emergency Services; New York State Energy Research and Development Authority. *Transporting Crude Oil in New York State: A Review of Incident Prevention and Response Capacity*; State of New York: Albany, NY, USA, 2014; pp. 8–110.

23. State of Oregon. *Preliminary Statewide Rail Safety Review*; Oregon State Legislature: Salem, OR, USA, 2014; pp. 1–78.

24. Margolis, J. *Runaway Risks, Oil Trains and Government's Failure to Protect People, Wildlife and the Environment*; Center for Biological Diversity: Portland, OR, USA, 2015; pp. 1–15.

25. Etkin, D.E.; Joeckel, J.; Walker, A.H.; Scholz, D.; Moore, C.; Baker, C.; Hatzenbuhler, D.; Patton, R.G.; Lyman, E. *Washington State 2014 Marine & Rail Oil Transportation Study: Appendix C: In Depth: Crude by Rail Emergency & Spill Response*; Spill Prevention, Preparedness & Response Program, Washington State Department of Ecology: Olympia, WA, USA, 2014; pp. 1–570.

26. Liu, X.; Rapik Saat, M.; Barkan, C.P.L. Analysis of Causes of Major Train Derailment and Their Effect on Accident on Accident Rates. *Transp. Res. Rec.* **2012**, *2289*, 154–163. [CrossRef]

27. Liu, X.; Rapik Saat, M.; Barkan, C.P.L. Safety Effectiveness of Integrated Risk Reduction Strategies for Rail Transport of Hazardous Materials. *Transp. Res. Rec.* **2013**, *2374*, 102–110. [CrossRef]

28. Alves, T.M.; Kokinou, E.; Zodiatis, G.; Lardner, R.; Panagiotakis, C.; Radhakrishnan, H. Modelling of oil spills in confined maritime basins: The case for early response in the Eastern Mediterranean Sea. *Environ. Pollut.* **2015**, *206*, 390–399. [CrossRef] [PubMed]

29. Alves, T.M.; Kokinou, E.; Zodiatis, G. A three-step model to assess shoreline and offshore susceptibility to oil spills: The South Aegean (Crete) as an analogue for confined marine basins. *Mar. Pollut. Bull.* **2014**, *86*, 443–457. [CrossRef] [PubMed]

30. National Oil and Hazardous Substance Pollution Contingency Plan. In *Code of Federal Regulations*; Part 300, Title 40; STATUTE: Washington, DC, USA, 1994.

31. Bigbie, J. Energy Renaissance Waterway Impact. *J. Saf. Secur. Sea* **2015**, *72*, 18–20.

32. NEREIDS. Scenario of the Crete Table Top exercise (July 2013). Available online: http://www.nereids.eu/site/en/index.php?file=scenarios (accessed on 10 February 2016).

33. Adler, E.; Inbar, M. Shoreline sensitivity to oil spills, the Mediterranean coast of Israel: Assessment and analysis. *Ocean Coast. Manag.* **2007**, *50*, 24–34. [CrossRef]

34. Aurand, D. The Application of Ecological Risk Assessment Principles to Dispersant Use Planning. *Spill Sci. Tech. Bull.* **1995**, *2*, 241–247. [CrossRef]

35. U.S. Environmental Protection Agency. *Guidelines for Ecological Risk Assessment*; Federal Register: Washington, DC, USA, 1998; pp. 26846–26924.

36. Walker, A.H.; Scholz, D.; Pond, R.G.; Aurand, D.V.; Clark, J.R. *Lessons Learned in Ecological Risk Assessment Planning Efforts*, Proceedings of the International Oil Spill Conference, Tampa, FL, USA, 26–29 March 2001; pp. 185–190.

37. Aurand, D.; Walko, L.; Pond, R. *Developing Consensus Ecological Risk Assessments: Environmental Protection In Oil Spill Response Planning A Guidebook*; United States Coast Guard: Washington, DC, USA, 2000; pp. 1–133.

38. Walker, A.H.; Scholz, D.; McPeek, M. *Consensus Ecological Risk Assessment of Potential Transportation-related Bakken and Dilbit Crude Oil Spills in the Delaware Bay Area: Comparative Evaluation of Response Actions*; United States Coast Guard: Washington, DC, USA, 2016.

39. Kraly, J.; Pond, R.; Walker, A.H.; Caplis, J.; Aurand, D.V.; Coelho, G.M.; Martin, B.; Sowby, M. *Ecological Risk Assessment Principles Applied to Oil Spill Response Planning*, Proceedings of the International Oil Spill Conference, Tampa, FL, USA, 26–29 March 2001; pp. 177–184.

40. Aurand, D.; Coelho, G.M.; Pond, R.G.; Martin, B.; Caplis, LCDR J.; Kraly, J.; Sowby, M.; Walker, A.H. *Results from Cooperative Ecological Risk Assessments for Oil Spill Response Planning in Galveston Bay, Texas and the San Francisco Bay Area, California*, Proceedings of International Oil Spill Conference, Sacramento, CA, USA, 26–29 March 2001; pp. 167–175.

41. Aurand, D.; Pond, R.; Coelho, G.; Cunningham, LCDR M.; Cocanaur, LCDR A.; Stevens, L. *The Use Consensus Ecological Risk Assessments to Evaluate Oil Spill Response Options: Lessons Learned From Workshops in Nine Different Locations*, Proceedings of International Oil Spill Conference, Miami Beach, FL, USA, 15–19 May 2005; pp. 379–386.

42. Gelzer, C.C. *MER Policy Letter 01–14; Endangered Species Act (ESA) and Essential Fish Habitat (ESF) Consultation Process Guidance*; United States Coast Guard: Washington, DC, USA, 2013; pp. 1–6.

43. U.S. Energy Information Administration. New EIA monthly data track crude oil movements by rail. Available online: https://www.eia.gov/todayinenergy/detail.cfm?id=20592#tabs_Slider-1 (accessed on 14 January 2016).

44. U.S. Energy Information Administration. New EIA monthly data track crude oil movements by rail. Available online: https://www.eia.gov/todayinenergy/detail.cfm?id=20592#tabs_Slider-5 (accessed on 14 January 2016).

45. Clark, K.E.; Niles, L.J.; Burger, J. Abundance and distribution of migrant shorebirds in Delaware Bay. *Condor* **1993**, *95*, 694–705. [CrossRef]

46. Kauffman, G.J. *Socioeconomic Value of the Delaware River Basin in Delaware, New Jersey, New York, and Pennsylvania*; Delaware River Basin Commission: West Trenton, NJ, USA, 2011.

47. Sayer, J.; Sunderland, T.; Ghazoul, J.; Pfund, J.-L.; Sheil, D.; Meijaard, E.; Venter, M.; Boedhihartono, A.K.; Day, M.; Garcia, C.; *et al.* Ten Principles for a Landscape Approach to Reconciling Agriculture, Conservation, and Other Competing Land Uses. *Proc. Natl. Acad. Sci. USA* **2013**, *110*, 8349–8356. [CrossRef] [PubMed]

48. U.S. Environmental Protection Agency. *Emerging Contaminants—Perfluorooctane Sulfonate (PFOS) and Perfluorooctanoic Acid (PFOA)*; EPA 505-F-14-001. United States Environmental Agency: Washington, DC, USA, 2014.

Journal of
Marine Science and Engineering

MDPI

Article

Effects of Exposure of Pink Shrimp, *Farfantepenaeus duorarum,* Larvae to Macondo Canyon 252 Crude Oil and the Corexit Dispersant

Susan Laramore [1,*], William Krebs [1,2,†] and Amber Garr [1,3,†]

[1]   Harbor Branch Oceanographic Institute at Florida Atlantic University, 5600 US 1 North Fort Pierce, FL 34946, USA
[2]   Colorado Catch, LLC., PO Box 210, Sanford, CO 81151, USA; wdkthev@gmail.com
[3]   Fountain Valley School of Colorado, 6155 Fountain Valley School Road, Colorado Springs, CO 80911, USA; agarr@fvs.edu
*   Correspondence: slaramo1@fau.edu; Tel.:+1-772-242-2525
†   These authors contributed equally to this work.

Academic Editor: Merv Fingas
Received: 1 January 2016; Accepted: 2 February 2016; Published: 8 March 2016

Abstract: The release of oil into the Gulf of Mexico (GOM) during the Deepwater Horizon event coincided with the white and pink shrimp spawning season. To determine the potential impact on shrimp larvae a series of static acute (24–96 h) toxicity studies with water accommodated fractions (WAFs) of Macondo Canyon (MC) 252 crude oil, the Corexit 9500A dispersant, and chemically enhanced WAFS (CEWAFs) were conducted with nauplii, zoea, mysid, and postlarval *Farfantepenaeus duorarum*. Median lethal concentrations (LC_{50}) were calculated and behavior responses (swimming, molting, light sensitivity) evaluated. Impacts were life stage dependent with zoea being the most sensitive. Behavioral responses for all stages, except postlarvae, occurred at below LC_{50} values. Dispersants had the greatest negative impact while WAFs had the least. No short-term effects (survival, growth) were noted for nauplii exposed to sub-lethal CEWAFs 39 days post-exposure. This study points to the importance of evaluating multiple life stages to assess population effects following contaminant exposure and further, that the use of dispersants as a method of oil removal increases oil toxicity.

Keywords: *Farfantepenaeus duorarum*; shrimp; DWH; MC252 crude oil; Corexit 9500A dispersant

1. Introduction

The Gulf of Mexico (GOM) has some of the most productive coastal bodies of water in the world, making it a major source for the U.S. seafood industry and the most economically important of all domestic commercial seafood harvesting sectors [1]. One of the most important GOM fisheries is the shrimp industry, extending from Brownsville, Texas to Key West, Florida. In 2010 the GOM provided 68% of U.S.-harvested shrimp with a total dockside value of $281 million [2]. The fishery consists of three major species: brown shrimp (*Farfantepenaeus aztecus*), pink shrimp (*Farfantepenaeus duorarum*), and white shrimp (*Litopenaeus setiferus*) [3]. Both the pink and white shrimp began migrating offshore to spawn in the spring, with continued spawning migration throughout the summer (pink shrimp) and fall (white shrimp), while the spawning season for brown shrimp is less defined in terms of season [4–6]. Fertilized eggs pass through nauplii, zoea, and mysis stages in offshore waters before migrating back to coastal estuaries as postlarvae within three to four weeks, throughout the spring and summer, dependent on species. On April 20, 2010 the Deepwater Horizon (DWH) oil platform exploded resulting in 200 million gallons of oil being released into the GOM until the well was capped on July 15 [7,8]. It is estimated that 100,000 km^2 of the GOM was affected by the spill, which coincided with the spring spawning season of a number of key GOM species, including shrimp [3–6,8]. In an effort to contain the spill and prevent the oil from reaching the shoreline, booms, skimmers, burning, direct recovery, and dispersants were used [7]. It has been calculated that 1.9 million gallons of dispersant (Corexit 9527 and Corexit 9500A) were used [9]. Dispersants do not remove oil from water but act to break the oil into smaller droplets that are more readily dispersed into the water column [10]. While dispersant use decreases the amount of surface oil lessening the amount of oil that reaches shorelines, the small dispersed droplets that remain in the water column are now made available to pelagic organisms that inhabit the water column [7].

Several studies have shown negative impacts of oil or dispersed oil exposure on various invertebrates, including mollusks [11–15], echinoderms [16–18], and crustaceans [11,13,19–22]. Other studies have focused on determining the effect of dispersants on marine organisms [10,23–26]. Most studies concentrate on one stage of development. Early life stages are typically more sensitive to pollutants than juveniles or adults and may be impacted at concentrations that, at least on the surface, do not cause acute mortality in juveniles or adults. Yet, in the long term survival may be impacted by behavioral modifications such as reduced activity that may affect predator avoidance and food intake [27–31]. The aim of our study was to determine what concentration of MC252 oil, Corexit dispersant and chemically dispersed MC252 would adversely affect the survival, development, and behavioral responses of the four major larval stages of shrimp (nauplii, zoea, mysis, and postlarvae). Behavioral responses included swimming activity, light response, feeding, and molting.

2. Experimental Section

2.1. Animals

Various life stages: nauplii (stage N_1, N_5), proto-zoea (stage Z_1, Z_3), mysis (stage M_1, M_2), and six-day-old post-larvae (Pl_6) of shrimp (*Farfantepenaeus duorarum*) were obtained from two commercial shrimp facilities in Florida (Scientific Associates, Indiantown and Pine Island Aquafarms, St. James City, FL, USA).

2.2. Solution Preparation

Oil and dispersant solutions for all experiments were prepared with MC252 oil (British Petroleum Company, BP PLC, London, UK) or Corexit 9500A dispersant (Nalco/Exxon Energy Chemicals, Sugarland, TX, USA). Solutions were prepared following CROSERF procedures [32,33]. Prior to solution preparation, crude oil was physically weathered in the lab for 24 h by placement of oil in a beaker on a stir plate and mixing with a magnetic stir bar in the dark in a chemical fume hood. Stock solutions of water accommodated fractions (WAFs) of crude oil (2 g L^{-1}), dispersant (2 g L^{-1}) and chemically enhanced WAFs (CEWAFs) (1:10 ratio) were prepared in 2 L flasks of filtered, UV treated seawater (28 ppt), covered, and mixed at moderate intensity (25% vortex) for 24 h. Stock solutions were allowed to settle for 3 h prior to preparation of working solutions.

2.3. PAH Analysis

Samples of oil and dispersed oil stocks (2 g L^{-1}) used in the acute toxicity experiments were preserved in glass jars with dichloromethane (1:10 v/v) and extracted using modified EPA method 3510C (Mote Marine Laboratory, Sarasota, FL). Polycyclic aromatic hydrocarbons (PAHs, parent compounds, and homologues) were analyzed using GC/MS (Agilent 7890A/5975C), modified EPA method 8260. Total petroleum hydrocarbons (TPH) *n*-C9 to *n*-C42 were analyzed using a GC with a flame ionization detector (FID, Agilent 7890A, Agilent Technologies Inc., Santa Clara, CA, USA).

2.4. Acute Toxicity Bioassays

2.4.1. Survival (Determination of LC_{50} Values)

Acute static toxicity tests (May 2011) were conducted with N_2, Z_1, and M_1 using nominal WAF concentrations of 0, 6.25, 12.5, 25, 50, and 100 mg L^{-1}, and for Pl_6 shrimp using nominal WAF concentrations of 0, 50, 100, 200, and 400 mg L^{-1}. CEWAF concentrations of 0, 6.25, 12.5, 25, 50, and 100 mg L^{-1}, and dispersant concentrations of 0, 1.25, 2.5, 5, 10, and 50 mg L^{-1} were used for all four life stages, with five replicates per treatment for each of the three solutions. Nauplii ($N = 15$) were placed in finger bowls containing 50 mL of the appropriate solution. All other life stages ($N = 15$ Z_1, $N = 12$ M_1, Pl_6) were placed in 1000 mL beakers containing 600 mL of the appropriate solution. All containers were placed in incubators (28 °C, 12:12 h light:dark cycle). Shrimp were fed once per day. Nauplii and zoea were fed a mixture of *Chaetocerous gracilis* and *Isochrysis galbana*, mysis were fed rotifers, and postlarvae were fed a pelleted diet (Shrimp PL 40-9, Zeigler Bro. Inc., Gardners, PA, USA). Survival was assessed at 24, 48, 72, and 96 h. Lethal concentrations (LC_{50}) were determined using the trimmed Spearman-Karber method (ToxCalcv5.0).

2.4.2. Behavioral Responses

Several experiments were conducted to evaluate behavioral responses. Activity level (swimming behavior) and molting frequency were evaluated for M_1 and Pl_6 stages for both WAF (0, 100, 200, 400, 800, and 1200 mg L^{-1}) and CEWAF (0, 6.25, 12.5, 25, 50, and 100 mg L^{-1}) exposures, with five replicates per treatment group, and 12 shrimp per replicate. Activity level was scored on a scale of 1–4: 1 = actively moving, 2 = moderately active, 3 = lethargic/moving appendages only, 4 = dead. Molting frequency was calculated as the percent of shrimp that molted compared to the total number of shrimp.

Subsequent behavioral response experiments were conducted for CEWAF exposures only for N_5, Z_1, Z_3, and M_2 stages. Concentrations used varied based on life stage evaluated, with five replicates per treatment group, 15 shrimp per replicate. CEWAF concentrations of 0, 6.25, 12.5, 25, 50, and 100 mg L^{-1} were used for nauplii and mysis stages, while concentrations were adjusted to 0, 3.125, 6.25, 12.5, and 25 mg L^{-1} for the more sensitive proto-zoeal stages as determined by the LC_{50} experiments. Behavioral parameters assessed included activity and molting as defined above, and feeding and photo-taxic response. Feeding was scored on a 1–4 scale: 1 = actively feeding (food in gut, fecal strands), 2 = 50% or less feeding, 3 = 25% or less feeding, 4 = 0% feeding. Photo-taxic response was evaluated by placing a light source to one side of the container and noting the proportion of shrimp that were attracted to the light (N_5, Z_1) or avoided the light (Z_3, M_2). Photo-taxic response was scored on a 1–3 scale: for N_5, Z_1—1 = actively moving towards light, 2 = sluggish response, 3 = no response; for Z_3, M_2—1 = actively moving away from light, 2 = slow avoidance response, 3 = no response. The proportion of shrimp that underwent metamorphosis to the next stage was also noted in these experiments.

2.5. Sub-Lethal Toxicity Bioassays

Approximately 10,000 *L. duorarum* nauplii were evenly divided between one of six 13 L buckets containing either filtered, UV treated HBOI salt well water ($N = 3$) or 23 mg L^{-1} CEWAF ($N = 3$). During the 24 h exposure, shrimp were fed *Isochrysis galbana* during experimental exposure at a rate of 15,000 (or $\times 10^3$) cells/mL. Surviving shrimp from both control buckets and treatment buckets were sieved, combined and then redistributed into one of four 400-L larval rearing tanks (two control, two treatment) containing filtered, UV treated HBOI salt well water. On day 1 (24 h exposure) and on alternate days, seven to eight shrimp were randomly removed from each of the four tanks (15 control, 15 exposed) for 39 days, collected and placed in vials containing 10% NBT formalin. After a 24 h fixation period, shrimp were placed in 70% ethanol, examined microscopically, and photographed (Infinity 2 digital camera, Luminera Co., Sachse, TX, USA). Developmental stage was recorded and length measurements averaged for each data point using Infinity Analyze (Luminera Co.).

3. Results

3.1. PAH Analysis

The total PAH level in the CEWAF stock solution (1429 µg L^{-1}) was three times greater than that of the WAF stock solution (452 µg L^{-1}) while the TPH level (62,613 µg L^{-1}) in the CEWAF solution was 25 times greater than that of the WAF stock solution (2467 µg L^{-1}) (Table 1). The predominant compound was napthalene, which made up 83.5% of the compounds in the WAF and 65% of the compounds in the CEWAF stock solution. Compounds containing three and four carbon rings (e.g., anthracene, fluorene, pyrene, chrysene, and phenanthrene) were approximately two times greater in the CEWAF compared to the WAF solution.

Table 1. Individual PAH and total TPH and PAH concentrations (µg L^{-1}) of 2 g L^{-1} stock solutions of water accommodated fractions (WAF) and chemically enhanced water accommodated fractions (CEWAF) used to prepare working solutions used in the acute toxicity experiments.

Target Compounds	C rings	2 ppt CEWAF µg L^{-1}	2ppt WAF µg L^{-1}
Napthalene (C0-C4)	2	925.96	377.66
Acenaphthylene	2	6.40	0.05
Acenaphthene	2	0.61	0.67
Fluorene (C0-C4)	3	102.92	14.65
Anthracene (C0-C4)	3	235.38	30.14
Phenanthrene	3	34.79	8.36
Fluoranthene	3	1.53	0.18
Chrysene (C0-C4)	4	32.6	4.02
Pyrene (C0-C4)	4	61.5	6.71
Benzo[A]anthracence	4	0.21	0.14
Napthobenzothiophene (C0-C4)	4	1.32	0.16
Dibenzothiophene (C0-C4)	5	5.47	5.6
Benzo[B]fluorene	5	0.72	0.09
Benzo[B]fluoranthene	5	0.00	0.07
Benzo[K]fluoranthene	5	0.43	0.00
Benzo[E]pyrene	5	0.61	0.11
Benzo[A]pyrene	5	0.00	0.02
Perylene	5	0.77	0.13
Dibenzo[A,H]anthracene	5	0.00	0.01
Indeno[1,2,3-Cd]pyrene	6	0.01	0.00
Benzo[G,H,I]perylene	6	0.00	0.02
Total PAH in µg L^{-1}		1428.64	451.92
Total Petroleum Hydrocarbon C9-C42 in µg L^{-1}		62,613.50	2466.57

3.2. Acute Toxicity Bioassays

3.2.1. Survival (Nominal LC_{50} Values)

Dispersants had the greatest impact on survival of all larval stages while WAFs had the least, with the proto-zoeal (Z_1) stage exhibiting the greatest sensitivity and the postlarval (Pl_6) stage the least sensitivity to all three contaminants (Table 2). The dispersant had the greatest impact on Z_1 shrimp (3.1 mg L^{-1}, LC_{50}, 24 h; 2.5 mg L^{-1} LC_{50}, 48 h; 100% mortality, 72 h), with all other stages having similar LC_{50} values at 24 h (21–33 mg L^{-1}) (Table 2). LC_{50} values continued to decrease for nauplii (N_2) and mysis (M_1) over time, but not for Pl_6 (22–28 mg L^{-1}). CEWAFs, likewise, had the greatest impact on Z_1 shrimp (15.4 mg L^{-1}, LC_{50}, 24 h; 100% mortality, 48 h), with all other stages having similar LC_{50} values at 24 h (81.5–100 mg L^{-1}) (Table 2). LC_{50} values continued to decrease for all stages over time, but less for Pl_6 (44 mg L^{-1}, LC_{50}, 96 h) than for M_1 (8.5 mg L^{-1}, LC_{50}, 96 h). WAFs had the least impact on all life stages, with Z_1 being the most sensitive (67.4 mg L^{-1}, LC_{50}, 24 h; 25.5 mg L^{-1} LC_{50}, 48 h; 100% mortality, 72 h) and Pl_6 the least, with no LC_{50} value determined for concentrations tested (>400 mg L^{-1}, 96 h) (Table 2).

Table 2. Lethal concentration (LC_{50}) values for shrimp exposed to oil (WAF), dispersant (Corexit 9500A) and oil/dispersant mixture (CEWAF) as determined by a trimmed Spearman-Karber method (ToxCalc). Reported values include nominal LC_{50} (95% CL) (mg L^{-1}) and corresponding PAH and TPH levels (μg L^{-1}). Non-determined values are indicated by ND; Non-calculated values are indicated by NC.

Time	WAF LC_{50}	PAH	TPH	CEWAF LC_{50}	PAH	TPH	Corexit LC_{50}
24 h							
Nauplii	>100	NC	NC	81.5 (75.3, 88.1)	58	2,551	33.3 (31.8, 34.9)
Zoea 1	67.4 (39, 100)	15	83	15.4 (11.6, 20.4)	11	470	3.1 (0.7, 13.7)
Mysis 1	>100	NC	NC	84.6 (74.9, 95.7)	60	2,649	20.9 (19.2, 22.7)
PL 6	>400	NC	NC	99.7 (77.3, 100)	71	3,121	28.4 (24.2, 33.3)
48 h							
Nauplii	>100	NC	NC	41.5 (36.3, 47.5)	30	1,299	18.6 (16.8, 20.5)
Zoea 1	25.5 (22.5, 28.9)	6	31	ND	NC	NC	<2.5
Mysis 1	>100	NC	NC	47.4 (41.3, 54.3)	34	1,484	18.3 (16.1, 20.8)
PL 6	>400	NC	NC	70.4 (56.4, 87.9)	50	2,204	26.5 (22.4, 31.3)
72 h							
Nauplii	ND	NC	NC	ND	NC	NC	ND
Zoea 1	21.2 (17.7, 25.5)	5	26	ND	NC	NC	ND
Mysis 1	>100	NC	NC	31.9 (28.6, 35.7)	23	999	8.3 (6.8, 10.1)
PL 6	>400	NC	NC	49.2 (40, 60.5)	35	1,002	22.4 (20.8, 23.9)
96 h							
Nauplii	ND	NC	NC	ND	NC	NC	ND
Zoea 1	23.3 (20.9, 26)	5	29	ND	NC	NC	ND
Mysis 1	29.7	7	37	8.5 (7.1, 10.1)	6	266	2.6 (2.2, 3.0)
PL 6	>400	NC	NC	44 (36.5, 53.2)	31	1,377	22.5 (21.4, 23.8)

3.2.2. Survival (Determined PAH and TPH LC_{50} Values)

PAH and TPH values for nominal CEWAFs and WAFs could only be compared at 24 h for Z_1 and 96 h for M_1 (Table 2). Toxicity of CEWAFs and WAFs were similarly toxic when PAH concentrations were compared, however WAFs were more toxic than CEWAFs when TPH concentrations were compared.

3.2.3. Behavioral Response—WAF & CEWAF (M_1, Pl_6)

Activity (swimming ability) was significantly decreased for M_1 exposed to CEWAF and WAF (Figures 1 and 2). CEWAFs decreased M_1 activity at 50 mg L^{-1} (36 μg L^{-1} PAH) at 24 h ($F_{5,24}$ = 103.6, $p < 0.0001$), 12.5 mg L^{-1} (18 μg L^{-1} PAH) at 48 and 72 h ($F_{5,24}$ = 25.93, $p < 0.0001$; $F_{5,24}$ = 26.53, $p < 0.0001$) and 6.25 mg L^{-1} (4 μg L^{-1} PAH) at 96 h (($F_{5,24}$ = 119.73, $p < 0.0001$ (Figure 1). WAFs decreased M_1 activity at 800 mg L^{-1} (181 μg L^{-1} PAH) at 24 h ($F_{5,24}$ = 28.11, $p < 0.0001$), 400 mg L^{-1} (90 μg L^{-1} PAH) at 48 h ($F_{5,24}$ = 9.62, $p < 0.0001$) and 100 mg L^{-1} (23 μg L^{-1} PAH) at 72 and 96 h ($F_{5,24}$ = 12.38, $p < 0.0001$; $F_{5,24}$ = 25.05, $p < 0.0001$) (Figure 2).

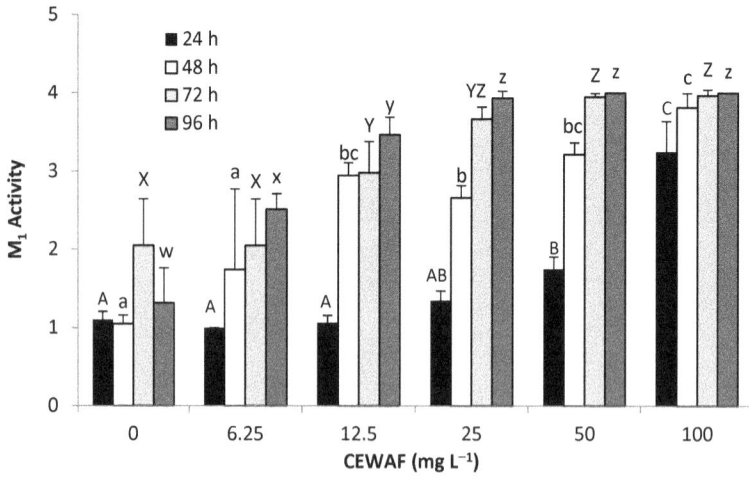

Figure 1. Average activity level (±S.D.) of *F. duorarum* mysis 1 (M₁) shrimp larvae exposed to chemically enhanced water accommodated fractions of MC252 crude oil (CEWAF). Treatment groups consisted of five replicates with 12 shrimp each: 1 = active, 2 = moderately active, 3 = lethargic, and 4 = dead. Numerical representations indicate statistical comparisons of exposure periods. Statistical differences were seen at all exposure times ($p < 0.0001$).

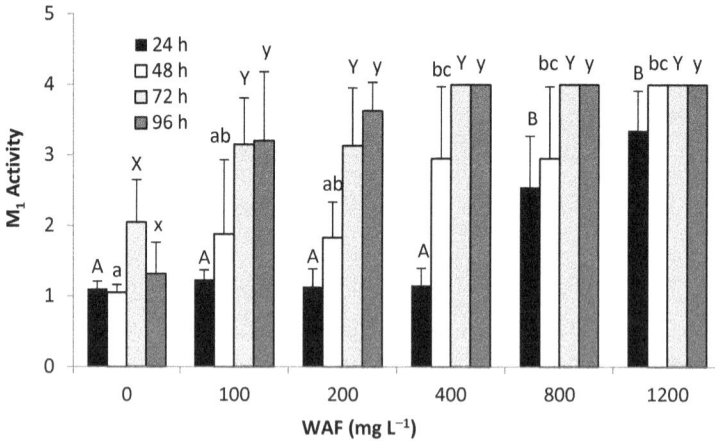

Figure 2. Average activity level (±S.D.) of *F. duorarum* mysis 1 (M₁) shrimp larvae exposed to water accommodated fractions of MC252 crude oil (WAF). Treatment groups consisted of five replicates with 12 shrimp each: 1 = active, 2 = moderately active, 3 = lethargic, and 4 = dead. Numerical representations indicate statistical comparisons of exposure periods. Statistical differences were seen at all exposure times ($p < 0.0001$).

Activity of Pl₆ was not affected by exposure to CEWAFs or WAFs at concentrations tested (Figures 3 and 4). There were significant differences in activity in Pl₆ exposed to CEWAFs at 48 h ($F_{5,24} = 4.56$, $p = 0.0046$) and 96 h ($F_{5,24} = 15.73$, $p = 0.0001$), however this was not dose dependent (Figure 3). There were significant differences in activity in Pl₆ exposed to WAFs at 48 h ($F_{5,24} = 3.81$, $p = 0.0111$), however activity was not dose dependent (Figure 4).

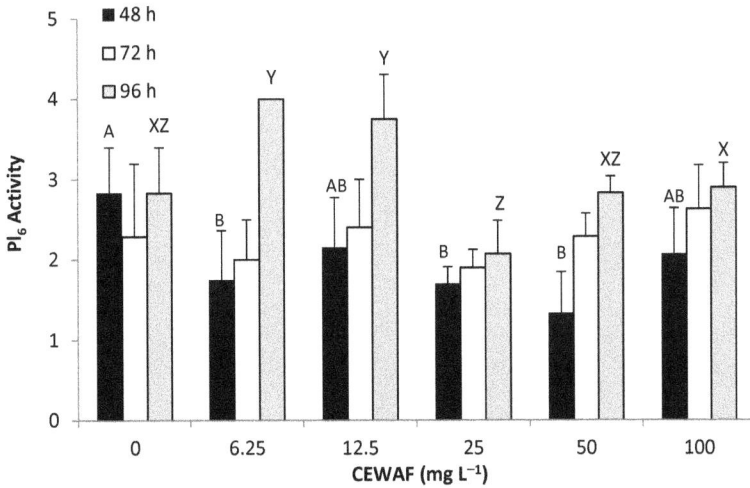

Figure 3. Average activity level (±S.D.) of *F. duorarum* postlarval (Pl$_6$) shrimp exposed to chemically enhanced water accommodated fractions of MC252 crude oil (CEWAF). Treatment groups consisted of five replicates with 12 shrimp each: 1 = active, 2 = moderately active, 3 = lethargic, and 4 = dead. Numerical representations indicate statistical comparisons of exposure periods ($p \leqslant 0.0046$, 48 h; $p = 0.3642$, 72 h; $p \leqslant 0001$, 96 h).

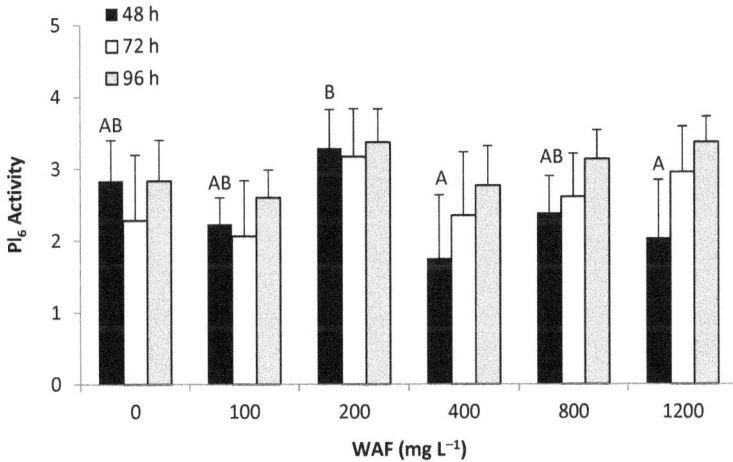

Figure 4. Average activity level (±S.D.) of *F. duorarum* postlarval (Pl$_6$) shrimp exposed to MC252 water accommodated fractions of crude oil (WAF). Treatment groups consisted of five replicates with 12 shrimp each: 1 = active, 2 = moderately active, 3 = lethargic, and 4 = dead. Significant differences were seen at 48 h ($p = 0.0111$).

CEWAFs $\geqslant$ 12.5 mg L^{-1} caused a significant increase in molting of M$_1$ (Figure 5). Significant differences were seen at both 24 ($F_{5,17} = 7.71$, $p = 0.0006$) and 48 h ($F_{5,17} = 63.79$, $p < 0.0001$).

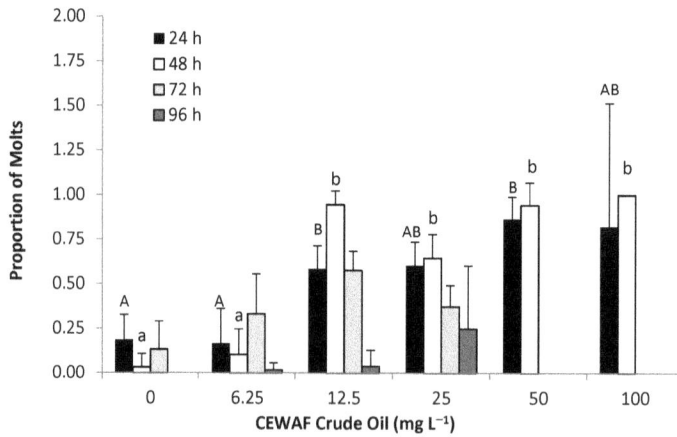

Figure 5. Proportion of *F. duorarum* M_1 shrimp larvae (±S.D.) that molted following exposure to CEWAFs. Treatment groups consisted of five replicates with 12 shrimp each. Numerical representations indicate statistical comparisons of exposure periods. Significant differences were seen at 24 and 48 h ($p \leqslant 0.0006$).

3.2.4. Behavioral Response—CEWAF (N_5, Z_1, Z_3, M_2)

Exposure of nauplii (N_5) to CEWAFs impacted swimming ability, feeding activity and phototaxic response (Table 3). Activity (swimming ability) of N_5 shrimp significantly decreased ($F_{4,25} = 98.8$, $p < 0.0001$) after 24 h exposure to all concentrations of CEWAF tested (12.5–100 mg L^{-1}) in a dose dependent matter. At 48 h, activity of N_5 shrimp was significantly different ($F_{4,25} = 98.8$, $p=.0262$) only at 100 mg L^{-1}. Feeding activity was likewise reduced ($F_{4,10} = 35.75$, $p < 0.0001$) at 24 h at all CEWAF concentrations. Phototaxtic response was also reduced at both 24 ($F_{4,25} = 81.45$, $p < 0.0001$) and 48 h ($F_{4,15} = 7.67$, $p = 0014$), in a dose dependent manner, with exposed shrimp, being slower to respond at both 24 and 48 h. There was no difference in metamorphosis from nauplii to zoea stages by 48 h, except for at 100 mg L^{-1}.

Table 3. Behavioral response of various larval stages of *F. duorarum* exposed to nominal CEWAF concentrations. Activity level is ranked on a 1–4 scale (1 = active, 4 = no response); feeding is ranked on a 1–4 scale (1 = 100% feeding, 4 = 0% feeding); phototaxic response is ranked on a 1–3 scale, which indicates attraction to light (1 = 100% attracted, 3 = 0% response) for nauplii (N_5) and zoea (Z_1), but light avoidance (1 = 100% avoidance, 3 = 0% response) for zoea (Z_3) and mysis (M_2). For each concentration a total of five replicates consisting of 15 shrimp each were averaged. Letters indicate significant differences between behavioral responses for concentrations at each time point.

Larval Stage	Time (h)	Concentration (mg L^{-1})	Activity Level (1–4) ± S.D.	Feeding (1–4) ± S.D.	% molts	Phototaxic (1–3) ± S.D.	Metamorphosis
N5	24	0	1.03±0.05 [a]	1.0 ± 0.0 [a]	-	1.02 ± 0.05 [a]	N5-Z1
		12.5	2.13 ± 0.05 [b]	3.33 ± 0.5 [b]	-	2.13 ± 0.05 [b]	N5-Z1
		25	2.35 ± 0.35 [bc]	3.67 ± 0.5 [b]	-	2.27 ± 0.41 [b]	N5-Z1
		50	2.28 ± 0.14 [bc]	4.0 ± 0.0 [b]	-	2.28 ± 0.14 [b]	N5-Z1
		100	2.98 ± 0.04 [c]	4.0 ± 0.0 [b]	-	3.0 ± 0.0 [c]	N5-Z1
	48	0	1.33 ± 0.52 [a]	-	-	1.5 ± 0.58 [a]	Z1-Z2
		12.5	2.17 ± 0.98 [a]	-	-	2.25 ± 0.29 [b]	Z1-Z2
		25	2.23 ± 0.74 [a]	-	-	2.34 ± 0.45 [bc]	Z1-Z2
		50	3.19 ± 0.95 [ab]	-	-	2.53 ± 0.67 [bc]	Z1-Z2
		100	3.83 ± 0.41 [b]	-	-	3.0 ± 0.0 [c]	Z1

Table 3. *Cont.*

Larval Stage	Time (h)	Concentration (mg L^{-1})	Activity Level (1–4) ± S.D.	Feeding (1–4) ± S.D.	% molts	Phototaxic (1–3) ± S.D.	Metamorphosis
Z1	24	0	1.67 ± 0.82 [a]	-	4%	1.67 ± 0.82 [a]	Z1-Z2 (4:1)
		3.125	2.29 ± 0.56 [ab]	-	12%	2.33 ± 0.58 [ab]	Z1
		6.25	2.54 ± 0.56 [b]	-	21%	2.63 ± 0.38 [b]	Z1
		12.5	2.92 ± 0.13 [bc]	-	21%	2.92 ± 0.13 [bc]	Z1
		25	3.0 ± 0.0 [c]	-	21%	3.0 ± 0.0 [c]	Z1
	48	0	2.0 ± 1.55	-	-	1.67 ± 1.03 [a]	Z1-Z2
		3.125	2.17 ± 1.47	-	-	1.83 ± 0.98 [a]	Z1-Z2
		6.25	3.0 ± 1.55	-	-	2.5 ± 0.77 [ab]	Z1-Z2
		12.5	3.17 ± 1.33	-	-	2.5 ± 0.84 [ab]	Z1-Z2
		25	4.0 ± 0.0	-	-	3.0 ± 0.0 [b]	-
	72	0	2.5 ± 1.64 [a]	-	-	2.0 ± 1.1	Z2
		3.125	2.75 ± 1.47 [a]	-	-	2.33 ± 1.03	Z1-Z2
		6.25	3.75 ± 0.61 [ab]	-	-	2.75 ± 0.61	Z1-Z2
		12.5	4.0 ± 0.0 [b]	-	-	3.0 ± 0.0	-
		25	4.0 ± 0.0 [b]	-	-	3.0 ± 0.0	-
Z3	24	0	1.0 ± 0.0 [a]	1.0 ± 0.0	0.80%	1.0 ± 0.0 [a]	-
		3.125	1.0 ± 0.0 [a]	1.0 ± 0.0	1.60%	1.0 ± 0.0 [a]	-
		6.25	1.0 ± 0.0 [a]	1.0 ± 0.0	0%	1.0 ± 0.0 [a]	-
		12.5	2.0 ± 0.0 [b]	1.0 ± 0.0	2.80%	3.0 ± 0.0 [b]	-
		25	2.0 ± 0.0 [b]	1.0 ± 0.0	1.60%	3.0 ± 0.0 [b]	-
	48	0	1.02 ± 0.04	1.0 ± 0.0	9.1%	-	Z3-M1 (3:2)
		3.125	1.0 ± 0.0	1.0 ± 0.0	15%	-	Z3-M1 (1:4)
		6.25	1.0 ± 0.0	1.0 ± 0.0	19.0%	-	Z3-M1 (1:4)
		12.5	1.0 ± 0.0	1.0 ± 0.0	13.3%	-	M1
		25	1.02 ± 0.04	1.0 ± 0.0	14.1%	-	Z3-M1 (2:3)
	72	0	1.13 ± 0.31	-	0%	1.29 ± 0.71 [ab]	-
		3.125	1.7 ± 1.1	-	0%	1.7 ± 1.1 [ab]	-
		6.25	1.03 ± 0.05	-	0%	1.03 ± 0.05 [a]	-
		12.5	1.12 ± 0.12	-	0%	2.39 ± 0.47 [b]	-
		25	1.07 ± 0.05	-	1.60%	2.03 ± 0.03 [b]	-
M2	24	0	1.0 ± 0.0 [a]	1.0 ± 0.0 [a]	4%	1.0 ± 0.0 [a]	M2
		12.5	1.2 ± 0.45 [a]	1.0 ± 0.0 [a]	1%	1.0 ± 0.0 [a]	M2
		25	1.84 ± 0.19 [b]	2.2 ± 0.45 [b]	28%	3.0 ± 0.0 [b]	M2
		50	1.9 ± 0.12 [b]	3.6 ± 0.55 [c]	37%	3.0 ± 0.0 [b]	M2
		100	2.0 ± 0.0 [b]	4.0 ± 0.0 [c]	51%	3.0 ± 0.0 [b]	M2
	48	0	1.0 ± 0.0 [a]	1.0 ± 0.0 [a]	2%	1.0 ± 0.0 [a]	-
		12.5	1.2 ± 0.45 [a]	1.0 ± 0.0 [a]	4%	2.0 ± 0.0 [b]	-
		25	1.35 ± 0.41 [a]	2.0 ± 0.0 [b]	28%	2.0 ± 0.0 [b]	-
		50	1.98 ± 0.08 [b]	2.6 ± 0.55 [bc]	2%	3.0 ± 0.0 [c]	-
		100	2.28 ± 0.04 [b]	3.4 ± 0.55 [c]	0%	3.0 ± 0.0 [c]	-
	72	0	1.8 ± 1.1 [a]	1.0 ± 0.0 [a]	0%	1.0 ± 0.0 [a]	M3
		12.5	2.29 ± 0.40 [a]	1.0 ± 0.0 [a]	0%	2.2 ± 0.45 [b]	-
		25	2.6 ± 0.55 [ab]	3.2 ± 1.1 [b]	3%	2.8 ± 0.45 [bc]	-
		50	3.0 ± 0.0 [b]	4.0 ± 0.0 [b]	0%	3.0 ± 0.0 [c]	-
		100	3.0 ± 0.0 [b]	4.0 ± 0.0 [b]	0%	3.0 ± 0.0 [c]	-

The behavior responses of proto-zoeal larvae (Z_1, Z_3) were impacted at lower concentrations of CEWAFs than were nauplii and mysis larvae (see below), and Z_1 larvae tended to exhibit these responses at levels lower than did Z_3 larvae (Table 3). Activity of Z_1 ($F_{4,25} = 6.68$, $p = 0.0008$) and Z_3 ($F_{4,25} = 707$, $p < 0.0001$) shrimp was significantly decreased at 6.125 and 12.5 mg L^{-1}, respectively, at 24 h. No significant differences were seen for either zoeal stage at 48 h, or for Z_3 shrimp at 72 h, however, Z_1 shrimp exposed to 12.5 and 25 mg L^{-1} for 72 h ($F_{4,25} = 2.96$, $p = 0.039$) were dead or moribund. Feeding activity was not monitored for Z_1 shrimp, however no difference was seen with Z_3 shrimp (Table 3). Molting frequency increased following exposure to 3.125 mg L^{-1} CEWAFs at 24 h for Z_1 and 48 h for Z_3 shrimp (Table 3). Phototaxic response was reduced for Z_1 shrimp at both 24 ($F_{4,25} = 7.78$, $p = 0.003$) and 48 h ($F_{4,25} = 2.77$, $p = 0.05$) at 6.125 and 12.5 mg L^{-1} respectively, and for Z_3 shrimp at 24 ($F_{4,25} = 3751$, $p < 0.0001$) and 72 h ($F_{4,25} = 5.12$, $p = 0.0036$) at 12.5 mg L^{-1}. All Z_1 control shrimp developed to Z_2 stage by 72 h, while some of the shrimp in all exposed groups were still in stage Z_1. An interesting pattern was seen in Z_3 exposed shrimp. A greater percentage of shrimp exposed to the highest CEWAF concentrations (12.5 and 25 mg L^{-1}) developed to M$_1$ stage than did control Z$_3$ shrimp, or shrimp exposed to lower CEWAF concentrations (Table 3).

Exposure of M$_2$ larvae to CEWAFs affected swimming ability, feeding response, phototaxic response and molting (Table 3). Swimming ability was affected at 25 mg L^{-1} at 24 h ($F_{4,20}$ = 21, p = 0.0007) and at 50 mg L^{-1} at 48 ($F_{4,20}$ = 13.8, p < 0.0001) and 72 h ($F_{4,20}$ = 3.0, p = 0.043). Feeding behavior was affected at 25 mg L^{-1} at all exposure times ($F_{4,20}$ = 99, p < 0.0001, 24 h; $F_{4,20}$ = 45, p < 0.0001, 48 h; $F_{4,20}$ = 48.9, p < 0.0001, 72 h) and shrimp exposed to higher concentrations had ceased feeding at 72 h. Molting frequency increased following exposure at 24 h for concentrations 25–100 mg L^{-1} (Table 3). Light avoidance was affected at 25 mg L^{-1} at 24 h ($F_{4,20}$ = 956, p < 0.0001) and at 12.5 mg L^{-1} at 48 ($F_{4,20}$ = 1483, p < 0.0001) and 72 h ($F_{4,20}$ = 44.2, p < 0.0001). No apparent lag in development to M$_3$ was noted between control and exposed groups.

3.3. Long Term Sublethal Effects

3.3.1. Survival

No difference in survival was seen between control and treatment groups. Survival was approximately 25% for both groups at day 39 post exposure.

3.3.2. Growth

No significant difference was seen in growth. Growth, as defined by total body length, was not significantly different between control and exposed groups from N$_5$ to Pl$_{28}$ ($F_{1,12}$ = 0.42, p = 0.5302) (Figure 6).

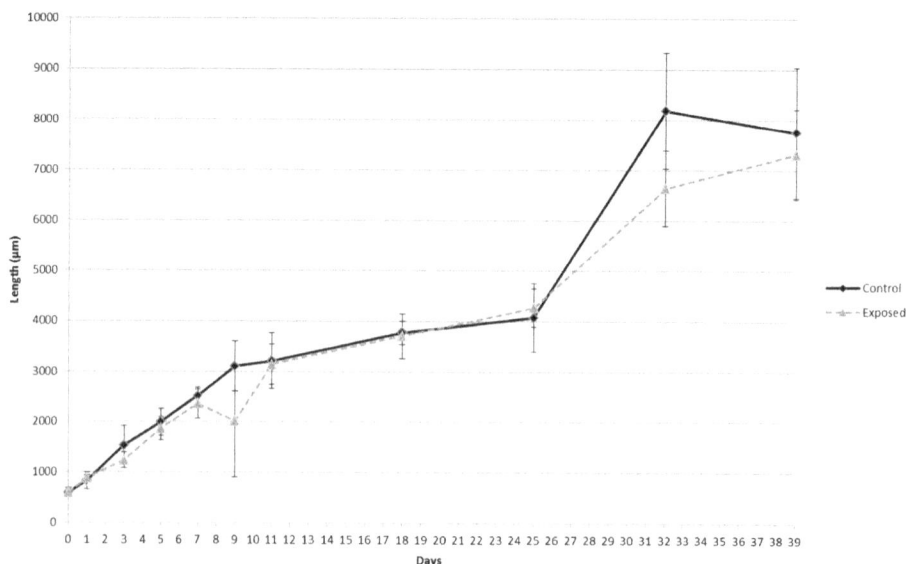

Figure 6. Average growth ($\pm$S.D.) of *F. duorarum* shrimp nauplii (N = 15) exposed to sub-lethal concentrations (23 mg L^{-1}) of CEWAFs for 24 h ($F_{1,12}$ = 0.42, p = 0.5302).

3.3.3. Developmental Stages

A slight developmental delay was seen between the control and exposed treatments on day five. Development from Z$_3$ to M$_1$ proceeded at a slower pace in the exposed groups resulting in a delayed development from M_2 to M_3 from day seven to nine. By day 11, development was similar and both groups reached Pl_1 (Figure 7).

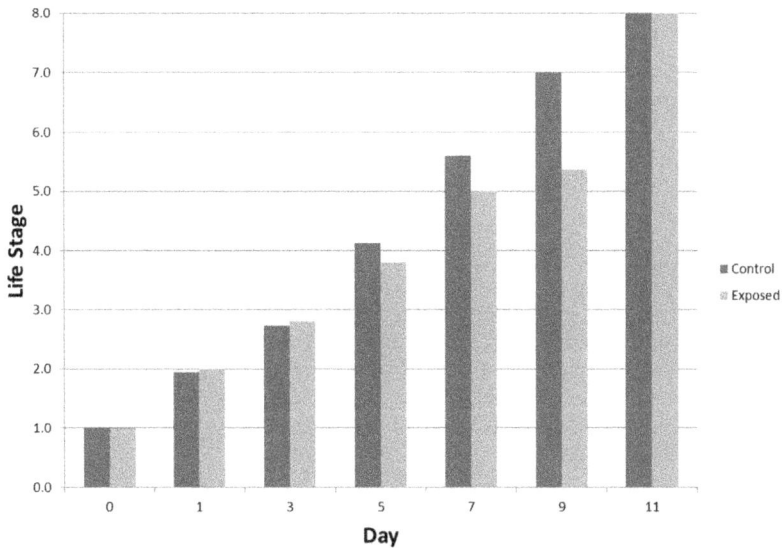

Figure 7. Development of *F. duorarum* nauplii ($N = 15$) exposed to sublethal concentrations of dispersed oil (23 mg L^{-1}) for 24 h. Life stage, (y axis), were assigned a numerical function for graphical representation. 1 = nauplii V, 2 = zoea 1, 3 = zoea 2, 4 = zoea 3, 5 = mysis 1, 6 = mysis 2, 7 = mysis 3, 8 = postlarvae.

4. Discussion

Exposure of various stages (nauplii, zoea, mysis, and postlarvae) of *F. duorarum* shrimp larvae to MC252 surrogate oil from the Deepwater Horizon well, and the primary dispersant, Corexit 9500A, used during the spill, adversely affected survival and behavior. Zoea (Z_1) were more sensitive to contaminant effects than other life stages. Dispersant exposure had a more pronounced affect, than did water accommodated fractions of crude oil (WAFs) or chemically enhanced WAFs (CEWAFs), and affected all larval stages equally and negatively. CEWAFs generally had a more negative impact than WAFs. Effects were dose and exposure dependent, with short-term sublethal effects resulting in slight developmental delays, with no longer term consequences to growth or survival seen in the laboratory.

The water column is the most likely route of contaminant uptake following an oil spill and therefore the toxicity of oil within the water column are commonly measured by analyzing the amount of oil contaminants within the WAFs and CEWAFs. Oil used was artificially weathered, to decrease the amount of volatile organic compounds (benzene, toluene, ethylbenzene, xylene) that tend to evaporate readily, in order to more closely mimic the type of oil that most organisms in the water column would likely encounter.

Concentrations of WAFs and CEWAFs used in this study represent moderate to maximum environmentally relevant levels based on reported literature. Oil levels ranging from 20 to 600 mg L^{-1} and dispersed oil levels ranging from 25 to 75 mg L^{-1} have been reported in the water column 24 h after a spill and may be a magnitude greater immediately following a spill [34]. Dispersants are used in relatively few spill events due in part to unfavorable conditions that are necessary for dispersants to work effectively [35]. Due to the magnitude of the Deepwater Horizon event large amounts of dispersant were used. Although the majority of water column organisms were likely exposed to either oil or dispersant and oil mixtures during the event it is possible that some organisms were inadvertently exposed to dispersants alone. Levels of dispersant used in these experiments were considered relevant based on reported literature. Dispersant concentrations of 0.1–15 mg L^{-1} have been reported in the field [10,31,36], although initial dispersant concentrations are generally below 10 mg L^{-1} (maximum range 5–15 mg L^{-1}), dropping to less than 1 mg L^{-1} in a few hours [24,36].

4.1. Acute Toxicity Effects

4.1.1. Survival

Survival is often used as the endpoint to determine oil toxicity effects. Previous studies have reported lethal concentration (LC_{50}) values for crustaceans, including shrimp, following acute exposure to oil, however, researchers typically focus on only one life stage and one contaminant (oil, dispersed oil, or dispersant). Early life stages tend to be more susceptible to toxic compounds than adults, which may negatively affect populations in affected areas, especially in the short-term. This study is unique in studying the effects of various shrimp larval stages simultaneously, to oil, dispersed oil, and a dispersant, and reporting behavioral responses along with LC_{50} values. We found that larval shrimp mortality varied dependent on developmental stage, and was not age dependent as zoea were more sensitive than nauplii. This is likely the result of differing feeding modalities at these two larval stages. Nauplii have undeveloped mouth parts and rely on their yolk sac for nutrition, while zoea are indiscriminate feeders and consume anything large enough to enter their mouth, and mysis seek out and capture their food [37]. Feed sources provided during this study varied: nauplii and zoea were provided with algae, mysis with rotifers, and postlarvae with commercial pellets. Exposure of WAFs or CEWAFs through either the water column or exposed feed resulted in similar alterations of metabolic enzymes in fish [38]. The addition of small quantities of feed may have resulted in larvae being exposed to oil contaminants through both the water column via the gills and the digestive tract via ingestion. We believed that administration of some feed was necessary to eliminate the likelihood of starvation as the cause of death as larvae, unlike postlarvae and juvenile shrimp, need to eat continuously. Preliminary experiments conducted with untreated and unfed larvae resulted in notable lethargy at 24 h and 50%–90% mortality of nauplii and zoea, respectively, at 48 h.

We noted that dispersant exposure negatively impacted all four larval stages at similar concentrations, although zoea were the most adversely affected, with all Z_1 shrimp dead by 48 h. Our reported values for 96 h exposures for *F. duorarum* M_1 and Pl_6 for Corexit 9500A, were similar to those previously reported for Corexit 9527 for *L. setiferus* postlarvae (96 h LC_{50}, 12–31 mg L^{-1}) [11,13]. Similar LC_{50} values (3.5–83 mg L^{-1}) have been reported following 48 to 96 h of exposure of other postlaval and juvenile crustaceans to Corexit 9500 [10,26,32,39,40].

Exposure to nominal concentrations of CEWAFs resulted in increased mortality compared to WAFs in our study. The majority of researchers have concluded that chemically dispersed oil is more toxic than physically dispersed oil [35]. However, reporting methods (nominal, PAH, TPH), may impact researchers conclusions [41]. In our study, when PAH, rather than nominal values were compared, the toxicity of WAFs and CEWAFs were equivalent, although WAFs appeared to be more toxic when TPHs were compared, similar to that seen for *L. setiferus* juveniles [13]. The increase in toxicity of CEWAFs is attributed to increased availability of PAHs in the water column through the creation of a large number of small oil droplets [35]. The PAH levels in the prepared CEWAF stocks in our study were three times greater and the TPH levels 25 times greater than in the WAF stocks. Oil droplets were observed in fecal strands and the digestive tract and fecal strands of some zoea and mysis larvae exposed to CEWAF concentrations $\geqslant$ 25 mg L^{-1} indicating ingestion of oiled particulates in larvae that were still feeding. At 50–100 mg L^{-1} oil was noted on appendages and molts of some shrimp larvae, implicating narcosis (PAH toxicity) and perhaps restricted motility as likely causes of mortality.

Although previous researchers have compared survival of crustaceans exposed to WAFs and CEWAFs by reporting LC_{50} values, there is little consistency in the reporting method (nominal, TPH,

PAH) which makes comparison difficult. In this study, we attempted to make cross-comparison easier by listing LC_{50} values for all three parameters. The majority of early crustacean studies were conducted with larval mysid shrimp, *Americanus* (*Mysiopsis*) *bahia*, where TPH values were reported, and 96 h LC_{50} values ranged from 0.15–83 mg L^{-1} WAFs and 0.5–120 mg L^{-1} CEWAFs [26,42,43]. That these results differ somewhat may be explained by variation in exposure methods used (constant, spiked, static renewal), however, each study reported a similar toxicity for WAFs and CEWAFs based on comparison of TPHs. Similar results have been reported for *Americamysis* (*Holmesimysis*) *costata* (1–35 mg L^{-1} WAF, 8–33 mg L^{-1} CEWAF, 96 h) and *L. setiferus* juveniles (6.5 mg L^{-1} WAFs, 5–7.5 mg L^{-1} CEWAFs, 96 h) [13,25,44]. In contrast, we report a 96 h TPH toxicity with larval *F. duorarum* (0.029–0.037 mg L^{-1} WAFs, 0.27–1.38 mg L^{-1} CEWAFs), indicating that WAFs were more toxic. This is in contrast to 96 h LC_{50} PAH values, in which little difference in toxicity was seen for WAFs and CEWAFs. In the wake of the Deepwater Horizon event, reported concentrations of TPAHs in May 2010 varied greatly dependent on site and sampling depth and ranged from 0 to 146 mg L^{-1} at the wellhead, 0 to 0.9 mg L^{-1} in field collected WAFs, 0 to 18 mg L^{-1} in field collected CEWAFs, and 0 to 0.17 µg L^{-1} in shoreline samples [45–47]. Following capping of the well in July concentrations in collected sample were significantly lower at all sites and depths.

4.1.2. Behavior

Factors in addition to mortality need to be considered when assessing contaminant effects, as behavioral responses, such as swimming ability, and response to stimuli affect the ability to locate prey or escape predation. Some researchers have reported behavioral inhibitory (IC_{50}) effects following exposure to lower levels of oil contaminants than at which mortality (LC_{50}) occurs. Changes in behavior due to sublethal exposures are considered to be the most sensitive indicators of environmental disturbance, and yet are among the least studied effects with regards to toxicity [48]. A variety of behavioral responses of marine organisms to pollutants, including oil, such as motivation (e.g., feeding response), sensory responses (e.g., phototaxis), and motor activity (e.g., swimming performance) are given in a summary of early work [49]. Examples of depressed feeding responses associated with PAHs have been shown in a variety of invertebrates including rotifers, crabs, and shrimp [50–52]. Examples of differential phototaxic responses associated with invertebrates have been reported with crabs and barnacles [27,53]. Invertebrates, such as crabs, shrimp, and barnacles, have also been shown to exhibit erratic swimming behavior in response to oil contaminants [27,53,54] and it has been postulated that differential sensory and motor responses that resulted in differential depth distribution might affect larval distribution and recruitment via directional current activity [53].

Some researchers have reported behavioral effects, such as reduction of swimming ability, at lower than LC_{50} concentrations [15,28]. Others have reported similar LC_{50} and EC_{50} values following exposure to oil contaminants, including swimming ability, settlement behavior and burying behavior [27,31,39]. Decreased swimming behavior is likely a result of narcosis typically seen in acute toxicity of high short-term exposures to naphthalene [55]. Narcotic chemicals affect the lipid bilayer in membranes reducing activity and the ability to react to stimuli, which may ultimately lead to mortality [56]. However narcosis does not account for other reported toxic effects such as deformities, edema, and cardiovascular effects [57] Regardless of whether this is the result of narcosis, or some other phenomenon the end result is that decreased swimming ability results in decreased ability to find food or escape predation, either of which will likely reduce survival.

In this study, swimming ability was significantly decreased for M_1 larvae exposed to both CEWAF and WAF at all exposure times. Exposure to CEWAFs caused an initial decrease in activity at concentrations that were two times less than LC_{50} values at 24 h and three to four less than LC_{50} values at 48 and 72 h. Similar results were seen with N_5, Z_1, Z_3, and M_2 larvae exposed to CEWAFs. However, results were stage dependent, in that Z_1 larvae were the most sensitive, whereas, older shrimp (Pl_6) did not exhibit reduced swimming ability at sublethal concentrations of either CEWAFs or WAFs. Although significant differences in activity in Pl_6 shrimp exposed to CEWAFs at 48 and

96 h and to WAFs at 48 h occurred, they were not dose dependent and thought to be due to water quality issues. Phototaxic response, whether attraction to light (N_5, Z_1) or avoidance (Z_3, M_2) followed a similar pattern to that seen with swimming ability, and this manifested itself in reduced feeding response for N_5, Z_1 stages.

4.1.3. Molting

Molting frequency increased in response to CEWAFs at 24 or 48 h post-exposure. Response was stage dependent with Z_1 and Z_3 larvae responding at lower concentrations than M_1 or M_2 larvae. It is postulated that this behavior is a stress response to compounds present in CEWAFs or an attempt by the shrimp to rid itself of oil adhering to the carapace or appendanges. Molting was not associated with metamorphosis to the next stage, as, except for the Z_3 stage, metamorphosis occurred at the same rate as the controls or was somewhat delayed, and control shrimp molted less frequently. An unwanted side effect of this response might be increased susceptibility to oil contaminants. Crustaceans are more susceptible to environmental stressors, including oil pollutants during molting, and may experience increased mortality [58,59]. In the present study, molting frequency was not followed after 72–96 h post-exposure. PAHs has been shown to increase the length of the intermolt period, resulting in decreased molting in a variety of invertebrates [58,60,61]. The use of additional measurements, such as biochemical endpoints, provide researchers with another set of tools for evaluating oil toxicity, allowing additional means of assessing the potential consequences of oil exposure for marine organisms, such as shrimp.

4.2. Sublethal Effects

Survival following sublethal exposures of invertebrates varies based on species and life stage [21,54,62,63]. In the present study, *F. duorarum* nauplii exposed to sub-lethal amounts (23 mg L^{-1}, LC_{10}) of CEWAFs for 24 h showed no difference in survival compared to controls over 39 days and was approximately 25% for both groups. Shrimp were cultured in larval tanks specific to the penaeid shrimp industry and industry operational procedures followed. Due to the sensitivity of handling the zoeal stage, tanks are not drained, water is instead added to partially filled tanks to alleviate water quality issues. Unfortunately, *Artemia* proliferated in the tanks, competing with shrimp for resources, and causing low survival in tanks regardless of exposure. Delayed morality has been reported by other researchers in shrimp and crab larvae and embryos exposed to low levels of WSF for short periods with zoea being more sensitive than later developmental stages [21,54,62]. Other research has shown no survival effects in crab zoea following either short term exposure or continuous exposure to low concentrations of oil [63].

Developmental delays have been reported for invertebrate larvae exposed to PAHs [21,63–65]. Developmental delays in invertebrates are typically accompanied by an increased period of intermolt following exposure to oil [58,64]. We saw a slight developmental lag in the exposed group between Z_3 and M_1 resulting in delayed development to subsequent mysis stages, but both groups developed in postlarvae at roughly the same time. Other researchers have reported similar findings. Zoeal stages of the mud crab *Rhithropanopeus harrisii* increased following prolonged exposure to chronic, low levels of WSF, however short term exposure had no impact [63,64]. Minor differences (one day lags) in development times were also reported in *Pandalus borealis* larvae in early stages of development but effects decreased at later stages [21]. The more time spent in pelagic larval stages, as would occur as a result of delayed development, may result in increased likelihood of predation, impact dispersion, increase time to maturity, and therefore negatively affect population growth rate [66].

In our study, despite slight developmental lags, no lasting growth effects were seen in the exposed group at day 39 (Pl_{28}). This is consistent with results reported in the literature for *Cancer irroratus* larvae exposed to WAFs, *R. harrisii* larvae exposed to low concentrations of naphthalene, and grass shrimp *Palaemonetes pugio* exposed to sublethal WAFs [59,67,68]. Other research has indicated decreased growth following exposure to oil. Decreased growth was reported in DWH exposed juvenile brown shrimp *Farfantepenaeus aztecus* but not in juvenile white shrimp *L. setiferus* exposed to the same waters [22].

5. Conclusions

This study shows that the concentrations of oil released and dispersant used during the DWH event could have negatively affected penaeid shrimp in the GOM, whether through altered behavioral responses, delayed development, or mortality. Even though the spill occurred during the spring spawning season and likely affected shrimp larvae at select locations, GOM shrimp populations as a whole do not appear to have been affected long term, perhaps in part due to fishery closures that were put in place following the spill [69].

Acknowledgments: This research was made possible by a grant from BP/The Gulf of Mexico Research Initiative through the Florida Institute of Oceanography #4710-1101-00B. We thank Sloane Wendell, BP GCRO Reference Material Account Manager for providing us with surrogate MC252 oil #SO-20110212-HMPA9-008; David Giddings, Strategic Technology Manager of Nalco Co., Sugarland Texas, for providing us the Corexit 9500A dispersant; and to FAU student Kaitlin Gallagher (currently UConn). This is HBOI-FAU publication # 1989.

Author Contributions: Susan Laramore and Amber Garr conceived and designed the experiments; Susan Laramore and William Krebs performed the experiments; Amber Garr analyzed the data; Susan Laramore and William Krebs wrote the paper; Amber Garr edited the paper.

Conflicts of Interest: The authors declare no conflict of interest.

References

1. National Marine Fisheries Service (NMFS). Fisheries of the United States 2003. In *National Marine Fisheries Service, Office of Science and Technology*; NMFS: Silver Spring, MD, USA, 2004.
2. National Marine Fisheries Service (NMFS). Fisheries of the United States 2010. In *National Marine Fisheries Service, Office of Science and Technology*; NMFS: Silver Spring, MD, USA, 2011.
3. Zimmerman, R.J.; Minello, T.J.; Rozas, L.P. Saltmarsh linkages to productivity of penaeid shrimps and blue crabs in the Northern Gulf of Mexico. In *Concepts and Controversies in Tidal Marsh Ecology*; Weinstein, M.P., Kreeger, D.A., Eds.; Kluwer Academic Publishers: Dordrecht, The Netherlands, 2000; pp. 293–314.
4. Linder, M.J.; Anderson, W.W. *Growth, Migration, Spawning and Size Distribution of Shrimp: Penaeus setiferus*; US Government Printing Office: Washington, DC, USA, 1956; Volume 56, pp. 555–645.
5. Renfro, W.C. *Life History Stages of Gulf of Mexico Brown Shrimp*; United States Bureau of Commercial Fisheries: Galveston, TX, USA, 1964; pp. 94–98.
6. Cummings, W.C. Maturation and spawning of the pink shrimp, *Penaeus duorarum* Burkenroad. *Trans. Am. Fish. Soc.* **1961**, *90*, 462–468. [CrossRef]
7. National Commission on the BP Deepwater Horizon Oil Spill and Offshore Drilling. Deep Water: The Gulf oil disaster and the future of offshore drilling. 2011; Report to the President. Available online: http://docs.lib.noaa.gov/noaa_documents/NOAA_related_docs/oil_spills/DWH_report-to-president.pdf (accessed on 16 November 2011).
8. Kelly, T.R. Environmental health insights into the 2010 Deepwater Horizon (BP) oil blowout. *Environ. Health Insights* **2010**, *4*, 61–63. [CrossRef]
9. Federal Interagency Solutions Group. Oil Budget Calculator: Deepwater Horizon, November 2010. Available online: http://www.restorethegulf.gov/sites/default/files/documents/pdf/OilBudgetCalc_Full_HQ-Print_111110.pdf (accessed on 16 November 2011).
10. George Aryes, A.; Clark, J.R. Aquatic toxicity of two Corexit dispersants. *Chemosphere* **2000**, *40*, 897–906. [CrossRef]
11. Fucik, K.W.; Carr, K.A.; Balcom, B.J. Toxicity of oil and dispersed oil to the eggs and larvae of seven marine fish and invertebrates from the Gulf of Mexico. In *The Use of Chemicals in Oil Spill Response*; ASTM STP 1252; Lane, P., Ed.; American Society for Testing and Materials: Philadelphia, PA, USA, 1995; pp. 135–171.
12. Le Hir, M.; Hily, C. First observations in a high rocky-shore community after the Erika oil spill (December 1999, Brittany, France). *Mar. Pollut. Bull.* **2002**, *44*, 1243–1252. [CrossRef]

13. Liu, B.; Romaire, R.P.; Delaune, R.D.; Lindau, C.W. Field investigation on the toxicity of Alaska North Slope crude oil (ANSC) and dispersed ANSC crude to Gulf killifish, Eastern oyster and white shrimp. *Chemosphere* **2006**, *62*, 520–526. [CrossRef] [PubMed]
14. Saco-Alvarez, L.; Bellas, J.; Nieto, O.; Bayona, J.M.; Albaiges, J.; Beiras, R. Toxicity and phototoxicity of water-accommodated fraction obtained from Prestige fuel oil and Marine fuel oil evaluated by marine bioassays. *Sci. Total Environ.* **2008**, *394*, 275–282. [CrossRef] [PubMed]
15. Laramore, S.; Krebs, W.; Garr, A. Effects of Macondo Canyon 252 oil (naturally and chemically dispersed) on larval *Crassostrea virginica* (Gmelin, 1791). *J. Shell. Res.* **2014**, *33*, 709–718. [CrossRef]
16. Temara, A.; Gulec, I.; Holdway, D.A. Oil induced disruption of foraging behavior of the asteroid keystone predator, *Coscinasterias muricata* (Echinodermata). *Mar. Biol.* **1999**, *133*, 501–507. [CrossRef]
17. Fernandez, N.; Cesar, A.; Salamanca, M.J.; DelValls, T.A. Toxicological characterization of the aqueous soluble phase of Prestige fuel oil using the sea urchin embryo bioassay. *Ecotoxicology* **2006**, *15*, 593–599. [CrossRef] [PubMed]
18. Beiras, R.; Saco Alvarez, L. Toxicity of seawater and sand affected by the Prestige fuel oil spill using bivalve and sea urchin embryogenesis bioassays. *Water Air Soil Poll.* **2006**, *177*, 457–466. [CrossRef]
19. Martínez-Jerónimo, F.; Villaseñor, R.; Ríos, G.; Espinosa-Chavez, F. Toxicity of the crude oil water soluble fraction and kaolin-adsorbed crude oil on *Daphnia magna* (Crustacea: Anomopoda). *Arch. Environ. Contam. Toxicol.* **2005**, *48*, 444–449. [CrossRef] [PubMed]
20. Roth, A.M.F.; Baltz, D.M. Short-term effects of an oil spill on marsh edge fishes and decapod crustacean. *Estuar. Coasts* **2009**, *32*, 565–572. [CrossRef]
21. Bechmann, R.K.; Larsen, B.K.; Taban, I.C.; Hellgren, L.I.; Moller, P.; Sanni, S. Chronic exposure of adults and embryos of *Pandalus borealis* to oil causes PAH accumulation, initiation of biomarker responses and an increase in larval mortality. *Mar. Poll. Bull.* **2010**, *60*, 2087–2098. [CrossRef] [PubMed]
22. Rozas, L.P.; Minello, T.J.; Miles, M.S. Effect of Deepwater Horizon oil on growth rates of juvenile penaeid shrimps. *Estuar. Coasts* **2014**, *37*, 1403–1414. [CrossRef]
23. Swedmark, M.; Granmo, A.; Kollberg, S. Effects of oil dispersants and oil emulsions on marine animals. *Water Res.* **1973**, *7*, 1649–1692. [CrossRef]
24. Wells, P.G. The toxicity of oil dispersants to marine organisms: A current perspective. In *Oil Spill Chemical Dispersants: Research, Experience and Recommendations*; ASTM International: Philadelphia, PA, USA, 1984; pp. 177–202.
25. Singer, M.M.; Aurand, D.; Coelho, G.; Bragin, G.E.; Clark, J.R.; Sowby, M.L.; Tjeerdema, R.S. Making, measuring and using water-accommodated fractions of petroleum for toxicity testing. In Proceedings of the 2001 International Oil Spill Conference, American Petroleum Institute, Tampa, FL, USA, 26–29 March 2001; pp. 1269–1274.
26. Fuller, C.; Bonner, J. Comparative toxicity of oil, dispersant and dispersed oil to Texas marine species. In Proceedings of the 2001 International Oil Spill Conference, American Petroleum Institute, Tampa, FL, USA, 26–29 March 2001; pp. 1243–1248.
27. Wu, R.S.S.; Lam, P.K.S.; Zou, B.S. Effects of two oil dispersants on phototaxis and swimming behavior of barnacle larvae. *Hydrobiologia* **1997**, *352*, 9–16. [CrossRef]
28. Barron, M.G.; Carl, M.G.; Short, J.W.; Rice, S.D. Photoenhanced toxicity of aqueous phase and chemically dispersed weathered Alaska North Slope crude oil to Pacific herring eggs and larvae. *Environ. Toxicol. Chem.* **2003**, *22*, 650–660. [CrossRef] [PubMed]
29. Georgiades, E.T.; Holdway, D.E.; Brennan, S.E.; Butty, J.S.; Temera, A. The impact of oil derived products on the behavior and biochemistry of the eleven armed asteroid *Coscinasterias muracata*, Echinodermata. *Mar. Environ. Res.* **2003**, *55*, 257–276. [CrossRef]
30. Lopez, B.A.; Lopez, D.A. Moulting frequency and behavioural responses to salinity and diesel oil in *Austromegabalanus psittacus* (Molina) (Cirripedia: Balanidae). *Mar. Freshw. Behav. Physiol.* **2005**, *38*, 249–258. [CrossRef]
31. Greco, G.; Corra, C.; Garaventa, F.; Chelossi, E.; Faimali, M. Standardization of laboratory bioassays with *Balanus amphitrite* larvae for preliminary oil dispersants toxicological characterization. *Chem. Ecol.* **2006**, *22*, 163–172. [CrossRef]

32. Clark, J.R.; Bragin, G.E.; Febbo, R.J.; Letinski, D.J. Toxicity of physically and chemically dispersed oils under continuous and environmentally realistic exposure conditions: Applicability to dispersant use decisions in spill response planning. In Proceedings of the 2001 International Oil Spill Conference, American Petroleum Institute, Tampa, FL, USA, 26–29 March 2001; pp. 1249–1255.

33. Barron, M.G.; Ka'aihue, L. Critical evaluation of CROSERF test methods for oil dispersant toxicity testing under subarctic conditions. *Mar. Pollut. Bull.* **2003**, *46*, 1191–1199. [CrossRef]

34. National Research Council (NRC). Committee on understanding oil spill dispersants. In *Oil Spill Dispersants Efficacy and Effects*; National Academies Press: Washington, DC, USA, 2005; p. 400.

35. Fincas, M. *A Review of Literature Related to Oil Spill Dispersants 1997–2008*; Prince William Sound Regional Citizens Advisory Council: Anchorage, AK, USA, 2008; p. 155.

36. Trudel, K. Environmental risks and trade-offs in Prince William Sound. In Proceedings of the Dispersant Application in Alaska: A Technical Update, Anchorage, AK, USA, 1998; pp. 159–188.

37. Treece, G.D.; Fox, J.M. *Design, Operation and Training Manual for an Intensive Culture Shrimp Hatchery*; Publication TAMU-SG-93–505; Texas A&M University, Sea Grant College Program: Galveeston, TX, USA, 1993; p. 187.

38. Cohen, A.M.; Gagnon, M.M.; Nugegoda, D. Alterations of metabolic enzymes in Australian bass, *Macquaria novemaculeata*, after exposure to petroleum hydrocarbons. *Arch. Environ. Contam. Toxicol.* **2005**, *49*, 200–205. [CrossRef] [PubMed]

39. Gulec, I.; Leonard, B.; Holdway, D.A. Oil and dispersed oil toxicity to amphipods and snails. *Spill Sci. Technol. Bull.* **1997**, *4*, 1–6. [CrossRef]

40. Fuller, C.; Bonner, J.; Page, C.; Ernest, A.; McDonald, T.; McDonald, S. Comparative toxicity of oil, dispersant and oil plus dispersant to several marine species. *Environ. Toxicol. Chem.* **2004**, *23*, 2941–2949. [CrossRef] [PubMed]

41. Perkins, R.A.; Rhoton, S.; Behr-Andres, C. Toxicity of dispersed and undispersed, fresh and weathered oil to larvae of a cold water species, Tanner crab (*C. bairdi*) and standard warm water test species. *Cold Reg. Sci. Technol.* **2003**, *36*, 129–140. [CrossRef]

42. Rhoton, S.; Perkins, R.A.; Braddock, J.F.; Behr-Andres, C. A cold weather species response to chemically dispersed fresh and weathered Alaska North Slope crude oil. In Proceedings of the 2001 International Oil Spill Conference, American Petroleum Institute, Tampa, FL, USA, 26–29 March 2001; pp. 1231–1236.

43. Wetzel, D.L.; Van Fleet, E.S. Cooperative studies on the toxicity of dispersants and dispersed oil to marine organisms: A 3 year Florida study. In Proceedings of the 2001 International Oil Spill Conference, American Petroleum Institute, Tampa, FL, USA, 26–29 March 2001; pp. 1237–1241.

44. Singer, M.M.; George, S.; Lee, I.; Jacobson, S.; Weetman, L.L.; Blondina, G.; Tjeerdema, R.S.; Aurand, D.; Sowby, M.L. Effects of dispersant treatment on the acute aquatic toxicity of petroleum hydrocarbons. *Environ. Contam. Toxicol.* **1998**, *34*, 177–187. [CrossRef]

45. Boehm, P.D.; Cook, L.L.; Murray, K.J. Aromatic hydrocarbon concentrations in seawater: Deepwater Horizon oil spill. In Proceedings of the 2011 International Oil Spill Conference, American Petroleum Institute, Portland, OR, USA, 23–26 May 2011; Volume 1, Abstract 371.

46. Echols, B.S.; Smith, A.J.; Gardinali, P.R.; Rand, G.M. Acute aquatic toxicity studies of Gulf of Mexico water samples collected following the Deepwater Horizon incident (2 May 2010 to 11 December 2010). *Chemosphere* **2015**, *120*, 131–137. [CrossRef] [PubMed]

47. Allan, S.E.; Smith, B.W.; Anderson, K.A. Impact of the Deepwater Horizon oil spill on bioavailable polycyclic aromatic hydrocarbons in Gulf of Mexico coastal waters. *Environ. Sci. Techno.* **2012**, *46*, 2033–2039. [CrossRef] [PubMed]

48. Gerhardt, A. Aquatic behavioral ecotoxicology—Prospects and limitations. *Hum. Ecol. Risk Assess.* **2007**, *13*, 481–491. [CrossRef]

49. Eisler, R.; Carney, G.C.; Lockwood, A.P.M.; Perkins, E.J. Behavioural responses of marine poikilotherms to pollutants (and discussion). *Philos. Trans. R. Soc. Lond. B Biol. Sci.* **1979**, *286*, 507–521. [CrossRef] [PubMed]

50. Takashashi, F.T.; Kittredge, J.S. Sublethal effects of the water soluble component of oil: chemical communication in the marine environment. In *The Microbial Degradation of Oil Pollutants*; Ahearn, D.G., and Mers, S.P., Eds.; Louisiana State University: Baton Rouge, LA, USA, 1973; pp. 259–264.

51. McCain, B.B.; Malins, D.C. Effects of petroleum hydrocarbons on selected demersal fishes and crustaceans. In *Sediment Toxicity Assessment*; Burton, G.A., Ed.; Lewis Publishers: Boca Raton, FL, USA, 1992; pp. 315–325.

52.   Jensen, L.K.; Carroll, J. Experimental studies of reproduction and feeding for two Artic dwelling *Calanus* species exposed to crude oil. *Aquatic. Biol.* **2010**, *10*, 261–271. [CrossRef]
53.   Bigford, TE. Effects of oil on behavioral response to light, pressure and gravity in larvae of the rock crab *Cancer irroratus*. *Mar. Biol.* **1977**, *43*, 137–148. [CrossRef]
54.   Brodersen, C. Rapid narcosis and delayed mortality in larvae of king crabs and kelp shrimp exposed to the water soluble fraction of crude oil. *Mar. Environ. Res.* **1987**, *22*, 233–239. [CrossRef]
55.   Rice, S.D.; Short, J.W.; Brodersen, C.C.; Mecklenburg, T.A.; Moles, D.A.; Misch, C.J.; Cheatham, D.L.; Karinen, J.F. Acute toxicity and uptake-depuration studies with Cook Inlet crude oil, Prudhoe Bay crude oil, No. 2 fuel oil, and several subarctic marine organisms. In *Northwest Fisheries Center Auke Bay Fisheries Laboratory, Processed Report*; National Marine Fisheries Service, NOAA: Juneau, AK, USA, 1976; p. 90.
56.   Van Wezel, A.P.; Opperhuizen, A. Narcosis due to environmental pollutants in aquatic organisms—Residue-based toxicity, mechanisms and membrane burdens. *Crit. Rev. Toxicol.* **1995**, *25*, 255–279. [CrossRef] [PubMed]
57.   Barron, M.G.; Carls, M.G.; Heintz, R.; Rice, S.D. Evaluation of fish early life stage toxicity models of chronic embryonic exposures to complex polycyclic aromatic hydrocarbon mixtures. *Toxicol. Sci.* **2004**, *78*, 60–67. [CrossRef] [PubMed]
58.   Weis, J.S.; Cristini, A.; Rao, K.R. Effects of pollutants on molting and regeneration in *Crustacea*. *Amer. Zool.* **1992**, *32*, 495–500. [CrossRef]
59.   Johns, D.M.; Pechenik, J.A. Influence of water accommodated fraction of No. 2 fuel oil on energetics of *Cancer irroratus* larvae. *Mar. Biol.* **1980**, *55*, 247–254. [CrossRef]
60.   Karinen, J.F.; Rice, S.D. Effects of Prudhoe bay crude oil on molting tanner crabs, *Chionoecetes bairdi*. *Mar. Fish. Rev.* **1974**, *36*, 31–37.
61.   Cantelmo, A.; Lazell, R.; Mantel, L. The effects of benzene on molting and limb regeneration in juvenile *Callinectes sapidus*. *Mar. Bio. Lett.* **1981**, *2*, 333–343.
62.   Cucci, T.L.; Epifanio, C.E. Long term effects of water soluble fractions of Kuwait crude oil on the larval and juvenile development of the mud crab *Eurypanopeus depressus*. *Mar. Biol.* **1979**, *55*, 215–220. [CrossRef]
63.   Laughlin, R.B., Jr.; Ng, J.; Guard, H.E. Hormesis: A response to low environmental concentrations of petroleum hydrocarbons. *Science* **1981**, *211*, 705–707. [CrossRef] [PubMed]
64.   Laughlin, R.B. Jr.; Young, L.G.L.; Neff, J.M. A long term study of the effects of water soluble fractions of No. 2 fuel oil on the survival, development rate and growth of the mud crab *Rhithropanopeus harrisii*. *Mar. Biol.* **1978**, *47*, 87–95. [CrossRef]
65.   Bang, H.W.; Lee, W.; Kwak, L.S. Detecting points as developmental delay based on the life-history development and urosome deformity of the harpacticoid copepod, *Tigriopus japonicus sensu lato*, following exposure to benzo(a)pyrene. *Chemosphere* **2009**, *76*, 1435–1439. [CrossRef] [PubMed]
66.   Worboys, M.A.; Leung, K.M.Y.; Grist, E.P.M.; Crane, M. Time should be considered in developmental ecotoxicity test. *Mar. Poll. Bull.* **2002**, *45*, 92–99. [CrossRef]
67.   Laughlin, R.B., Jr.; Neff, J.M. Ontogeny of respiratory and growth responses of larval mud crabs *Rhithropanopeus harrisii* exposed to different temperatures, salinities and naphthalene concentrations. *Mar. Ecol. Prog. Ser.* **1981**, *5*, 319–332. [CrossRef]
68.   Gunderson, D.T.; Kristanto, S.W.; Curtis, L.R.; AL-Yakoob, S.N.; Metwally, M.; Al-Aljimi, D. Subacute toxicity of the water soluble fractions of Kuwait crude oil and partially combusted crude oil on *Menidia beryllina* and *Palaemontes pugio*. *Arch. Environ. Contam. Toxicol.* **1996**, *31*, 1–8. [CrossRef]
69.   Van der Ham, J.L.; de Musert, K. Abundance and size of Gulf shrimp in Louisiana's coastal areas following the Deepwater Horizon oil spill. *PLOS ONE* **2014**, *9*, e108884. [CrossRef]

Journal of
*Marine Science
and Engineering*

MDPI

Article

Vegetation Impact and Recovery from Oil-Induced Stress on Three Ecologically Distinct Wetland Sites in the Gulf of Mexico

Kristen Shapiro *, Shruti Khanna and Susan L. Ustin

Department of Land, Air, and Water Resources, University of California, Davis, CA 95616, USA;
shrkhanna@ucdavis.edu (S.K.); slustin@ucdavis.edu (S.L.U.)
* Correspondence: kdshapiro@ucdavis.edu; Tel.: +1-510-579-1597

Academic Editor: Magnus Wahlberg
Received: 23 December 2015; Accepted: 21 April 2016; Published: 3 May 2016

Abstract: April 20, 2010 marked the start of the British Petroleum Deepwater Horizon oil spill, the largest marine oil spill in US history, which contaminated coastal wetland ecosystems across the northern Gulf of Mexico. We used hyperspectral data from 2010 and 2011 to compare the impact of oil contamination and recovery of coastal wetland vegetation across three ecologically diverse sites: Barataria Bay (saltmarsh), East Bird's Foot (intermediate/freshwater marsh), and Chandeleur Islands (mangrove-cordgrass barrier islands). Oil impact was measured by comparing wetland pixels along oiled and oil-free shorelines using various spectral indices. We show that the Chandeleur Islands were the most vulnerable to oiling, Barataria Bay had a small but widespread and significant impact, and East Bird's Foot had negligible impact. A year later, the Chandeleur Islands showed the strongest signs of recovery, Barataria Bay had a moderate recovery, and East Bird's Foot had only a slight increase in vegetation. Our results indicate that the recovery was at least partially related to the magnitude of the impact such that greater recovery occurred at sites that had greater impact.

Keywords: oil spill; marshes; AVIRIS; spectroscopy; remote sensing; oil impact; cordgrass; mangroves; Mississippi Deltaic Plain; Gulf of Mexico

1. Introduction

The British Petroleum Deepwater Horizon (BP-DWH) oil spill was the largest marine oil spill in US history [1]. The spill deposited oil along the shoreline of the Mississippi Deltaic Plain (MDP) which is home to the largest expanse of contiguous wetlands in the United States [2]. Wetlands offer a variety of ecological services including water purification, carbon storage, storm protection and nutrient cycling [3]. The MDP also supports one of the world's largest petroleum development infrastructures and is subject to chronic low-level oil spills [3]. The coastal wetlands and barrier islands of the MDP are already experiencing extreme land loss at the rate of ~43 km^2/year [4]. Oil contamination can escalate the rate of land loss through reduction of vegetation cover and loss of live plant roots which stabilize the soil [5].

The physical effects of oil on plant health are varied and based on a number factors including plant type, oil type, oil concentration, persistence, and extent of oil penetration into the marsh [6]. Plant mortality is higher when oil coats the leaf surfaces leading to reduced gas exchange via the stomata [7]. This results in oxygen-stressed roots (or pneumatophores in the case of mangroves) that further reduce plant growth and water uptake for transpiration resulting in high temperature stress. Plant roots are further stressed when oil coats the soil resulting in reduced soil conditions [7]. These stresses are manifested through leaf chlorosis and stem discoloration as plants lose chlorophyll [8]. Changes in plant health can be measured remotely by looking at the spectral response of vegetation.

Healthy plants reflect strongly in the near-infrared region of solar radiation and absorb strongly in the visible region due to photosynthetic pigments. The sharp increase in leaf reflectance between 680 and 750 nm is called the "red edge". As plants lose pigments due to chlorosis, the red edge becomes less sharp, reflectance in the visible region rises, and reflectance in the near-infrared (NIR) drops [9]. Thus plant chlorosis can be measured using spectral pigment indices such as the Normalized Difference Vegetation Index (NDVI) and modified NDVI (mNDVI) that track changes in chlorophyll content [10,11]. Furthermore, the red edge can be used as an indicator of plant stress based on a shift in its wavelength position and a decline in its slope [8,9].

When plants experience stress, they may also start to lose water, which manifests as a lower absorption in the shortwave-infrared region (1500–2500 nm) [8,12]. Water loss can be measured using spectral indices such as Normalized Difference Infrared Index (NDII), Absorption Depth of Water at 980 nm (ADW1) and ADW at 1240 nm (ADW2) [12,13]. As vegetation senesces, angle indices like Angle at NIR (ANIR) and Angle at Red (ARed) can track landcover change from healthy green vegetation to non-photosynthetic vegetation (NPV) and eventually soil [14].

Remote sensing for monitoring spatially extensive disasters such as the BP-DWH oil spill is important since field surveys are often costly and time consuming, yet may still fail to cover the entire impacted area. Furthermore, sensitive wetland ecosystems that have been stressed by oil contamination can be further damaged by site visits and remediation activities [15,16]. Remote sensing offers an alternative to extensive field surveys by providing 100% sampling of the affected area, as well as repeat monitoring that can extend over multiple years to track recovery in addition to impact. A handful of remote sensing studies have mapped plant stress in response to oil contamination [8,13,17,18]. A popular method for tracking oil-induced stress has been to identify a shift in the red edge towards longer [8,19,20] or shorter wavelengths [21]. However, Khanna *et al.* [22] successfully used pigment and plant water spectral indices to elucidate patterns of oil impact on saltmarsh vegetation in Barataria Bay, Louisiana.

The objective of this study is to compare the impact of the BP-DWH oil spill on wetland vegetation and monitor recovery in the following year across three diverse ecosystems in the MDP. Several studies have documented the impact from the BP spill for the gulf wetlands, but the majority of these studies have focused on a single marsh vegetation species or a single site in the gulf [5,17,23–25]. We wanted to compare the effect of the BP-DWH oil spill across multiple ecosystems in the gulf as well as evaluate recovery of live foliar canopy a year later. We used hyperspectral imagery from the fall of 2010 to assess the impact of oil-induced stress on wetland vegetation and compared it to imagery from 2011 to look for evidence of recovery from the oil. To evaluate oil impacts on the marsh ecosystem, we used several vegetation indices as measures of plant stress. For recovery, we used change detection of classified images between years to see if the area of green vegetation had increased or decreased.

2. Data and Methods

2.1. Study Sites

The MDP estuary complex contains the largest saltmarsh area and the second largest fresh marsh area in the continental United States [26]. The MDP also contains barrier islands that support black mangrove shrublands and serve a protective role for the mainland by absorbing the brunt of tidal surges from incoming storms [27]. Understanding the role of vegetation and its response to a contamination event is important for effective management in the face of continued land loss as well as effective recovery in times of disaster [28]. Hence we chose three sites for our study: Barataria Bay (saltmarsh), East Bird's Foot (freshwater and intermediate marsh), and the Chandeleur Islands (mangrove-cordgrass marshland) that represent three distinct ecosystems of the MDP (Figure 1).

Figure 1. Location of all three study sites in the Gulf of Mexico.

Barataria Bay (BB) is located approximately 160 km from the spill site in an interlobe basin between the current Bird's Foot delta and the abandoned Lafourche delta lobes [29]. BB consists solely of saltmarshes as it no longer receives significant fresh water or sediment input due to the levees along the Mississippi River and the closure of Bayou Lafourche. The dominant vegetation in the low intertidal saltmarshes is *Spartina alterniflora* (saltmarsh cordgrass) and *Juncus roemerianus* (needlegrass rush), with subdominants *Spartina patens* (salt meadow cordgrass), *Distichlis spicata* (saltgrass) and *Batis maritima* (saltwort) more common in the higher marsh [3]. Since the BP-DWH oil spill occurred offshore, the oil came in with the tides and primarily contaminated the seaward edges of these marshes. Our field data sources (Section 2.2) indicate that the marsh edge vegetation in BB is dominated by *S. alterniflora* followed by *J. roemerianus*.

Our second study site is on the Chandeleur Islands (CI), a 72 km long barrier island chain located approximately 140 km from the spill site. The islands form a diverse landscape of beaches, dunes, and marshes. The Chandeleur Islands are frequently overwashed due to tidal surges during tropical and temperate storms. As a result, the vegetation plays an important role in stabilization and sand deposition [27]. CI is a remote area and studies in this area are rare, even in the aftermath of the BP-DWH spill. However, based on the Natural Resource Damage Assessment (NRDA) data [30], our field data, and high resolution aerial imagery (AeroMetric), we determined that our study area is dominated by *Avicennia germinans* (black mangrove) and *Spartina alterniflora* with subdominants *Salicornia* spp. (pickleweed) and *Distichlis spicata* (seashore saltgrass).

The East Bird's Foot (EBF) site is located in the current primary delta lobe approximately 80 km from the spill site. This area continues to receive fresh water and sediment from the Mississippi River, and contains approximately 61,650 acres of wetlands, 81% of which are fresh water, 17% are intermediate, and 2% are brackish or saline [31]. The dominant vegetation for intermediate marshes is *Spartina patens*, *Phragmites australis* (common reed), *Sagittaria falcata* (bull-tongue arrowhead), and *Alternanthera philoxeroides* (alligator weed) [3,32]. The dominant vegetation in freshwater marshes is *Panicum hemitomon* (maiden cane), *Sagittaria falcata*, *Eleoharis* spp. (spike rush), and *Alternanthera philoxeroides* [3,32]. The area of EBF most affected by oil and present in our image data was almost exclusively dominated by *Phragmites australis* [32–34].

2.2. Field Data

For the selection of oiled and oil-free areas at each site, we used shoreline data collected from May through September of 2010 provided by the Shoreline Cleanup Assessment Technique (SCAT) program through the Environmental Response Management Application (ERMA), part of the National Oceanic and Atmospheric Administration (NOAA). The SCAT data is a collection of multiple ground surveys starting from May 2010 which contain mapped shorelines based on observed presence of oil. The shorelines are assigned a level of oil contamination based on four broad descriptive categories: Heavy, Moderate, Light, and No Oil [33].

For information on plant species affected by oil, we also relied on field data collected by David Baker of Tulane University in all three study areas between 2012 and 2014 as part of this study. 10 m × 10 m plots were collected along 100 m transects that started at the water's edge and moved inland perpendicularly to the shore. Approximately eight points were collected per transect spread evenly along its length with 46 transects collected. Species composition, cover, and degree of oiling were noted for each plot. This data was further supplemented with vegetation composition data from the Coastwide Reference Monitoring System (CRMS) [32,35] and the Natural Resource Damage Assessment (NRDA) [30].

2.3. Image Data and Preprocessing

To test the impact of oil on the wetlands in 2010, we used AVIRIS imagery from September (for BB and CI) and October (for EBF) of 2010 (Table 1). According to NOAA tide tables (http://tidesandcurrents.noaa.gov/), tide levels were only different between imagery collection dates for Chandeleur Islands (Table 1). This was taken into account when interpreting the results for Chandeleur Islands.

Table 1. Site information regarding data acquisition for each study site (East Bird's Foot—EBF, Barataria Bay—BB, and Chandeleur Islands—CI) such as time and date of acquisition, pixel resolution, number of flightlines, and tidal stage. Water levels are calculated based on the mean lower low water (MLLW) height datum.

Site	Year	Flight Dates	Pixel Resolution (m)	# Flightlines	Water Level (m)	Shoreline Analyzed (km)
BB	2010	09/14	3.5 × 3.5	4	0.21	30.4
	2011	08/25	7.7 × 7.7	2	0.25	
CI	2010	09/21	3.4 × 3.4	2	0.66	3.2
	2011	08/11	7.7 × 7.7	2	0.25	
EBF	2010	10/03	3.4 × 3.4	2	0.11	4.3
	2011	10/14	3.4 × 3.4	2	0.18	

To test recovery in 2011, a year after the oil spill ended, we acquired AVIRIS data over the same sites in August 2011 (For BB and CI) and in October 2011 (for EBF). While the imagery for BB and CI was flown a month earlier in 2011, these date differences are not likely to affect the results as the growing conditions were similar for the months in which the data was collected. In 2010 and 2011, both sites received substantial rainfall the month preceding the imagery and very little rainfall in the month when the imagery was acquired. The average temperature for both sites ranged from 27–31 °C during September 2010 and August 2011. Both the 2010 and 2011 data were atmospherically calibrated, georectified, co-registered [13] and resampled to the 2011 spatial resolution (7.7 m for BB and CI and 3.4 m for EBF). We calculated spectral indices (Table 2) for each site in 2010.

Table 2. Seven indices chosen to study severity of oil impact. R_R, R_{NIR} and R_{SWIR} are the reflectance values in the Red, Near-Infrared (NIR), and Shortwave-Infrared (SWIR) bands, respectively, and λ_R, λ_{NIR}, and λ_{SWIR} are the center wavelengths for the Red, NIR and SWIR bands.

Acronym	Formula	Plant	References
NDVI	$(R_{NIR} - R_R)/(R_{NIR} + R_R)$	chlorophyll content and/or leaf area of the plant	[11]
mNDVI	$(R_{750} - R_{700})/(R_{750} + R_{700})$	chlorophyll content and/or leaf area of the plant	[10]
NDII	$(R_{NIR} - R_{923})/(R_{NIR} + R_{923})$	Water content	[12]
ANIR	Angle between (R_R, λ_R), (R_{NIR}, λ_{NIR}), and $(R_{SWIR}, \lambda_{SWIR})$	Phenology and stress	[14,36]
ARed	Angle between (R_R, λ_R), (R_R, λ_R), and (R_{NIR}, λ_{NIR})	Phenology and stress	[13,36]
ADW1	$0.5 \times (R_{1070} + R_{890}) - R_{990}$	Water content	[13]
ADW2	$0.5 \times (R_{1270} + R_{1070}) - R_{1167}$	Water content	[13]

Although we had access to high spatial resolution aerial imagery, QuickBird imagery, and WorldView-2 imagery for some of these sites, we chose the AVIRIS imagery for our comparative analysis because the spectral range (350–2500 nm) and spectral resolution (5–10 nm bandwidth) required to calculate our suite of indices were not available with any other imagery.

We also classified the AVIRIS images from both years into five classes: water, green vegetation, NPV, soil, and submerged aquatic vegetation (SAV) using a decision tree classifier [13]. The NPV class contains dead, dying, and senescent vegetation. We combined the NPV and soil classes into one combined NPV-Soil group for the analysis. The decision tree method applied in this study uses indices to differentiate classes that take advantage of their biophysical differences [22,37]. The index thresholds that provided maximum differentiation between classes were selected based on ANOVA tests and then used to build a binary decision tree. The decision tree method uses inputs derived from multiple binary nodes that are based on the characteristics of the dataset. The user builds the decision tree and chooses the threshold values that are used at each node in the tree. At each node, the classifier splits the data into one of two possible classes or groups. This method is analogous to using a plant identification key, such as the Jepson Manual of Higher Plants of California [38].

2.4. Image Analysis

2.4.1. Selection of Oiled and Oil-Free Shores

Oil first appeared on the shores of EBF and CI in the last week of April and first week of May 2010 [39,40]. However, we were unable to directly map the presence of oil on the marsh surface in the images acquired over EBF and CI in September–October. Two factors may have impacted our ability to map oil on the surface. Firstly, oil might have been present under the plant canopy obstructing its spectral signal. Secondly, plant water absorption features in the spectrum overlap with oil absorptions, making it difficult to detect oil even if it is on the surface of the plants.

On the other hand, oil reached the shores of BB in mid-July, and continued to arrive through August–September [41–44]. During the time the imagery was acquired, it was still fresh and coated the soil surface (Khanna, personal observation). Therefore, we were able to accurately classify oiled soil and NPV pixels for the September 19, 2010 image data (overall accuracy: 95%, Kappa: 0.88, $n = 40$) [13].

Since we were only able to map actual oil contamination at the BB site, we used the publically available SCAT data of oiled shorelines, recorded by visual observation during the period of the oil spill. We compared SCAT-observed shorelines to our classified oiled pixels in BB (Figure 2). We found that, of the three levels of oiling in the SCAT database for BB, the "Moderate" and "Light" oiled shorelines did not consistently correspond to regions of oiled pixels in our BB imagery. Figure 2c,d show examples of moderately oiled and lightly oiled shorelines. Moderately oiled shorelines had fragmented sections of oiled pixels, and lightly oiled shorelines had very few oiled pixels. We limited our analysis to "Heavily" oiled shorelines, comparing them to oil-free shorelines at all three sites, because the presence of oil-free regions within "Moderate" and "Light" oiled SCAT shorelines would weaken the spectral signal of the oil impacts (making it harder to detect). Shorelines selected for analysis in BB, CI and EBF are shown in Figures 2–4.

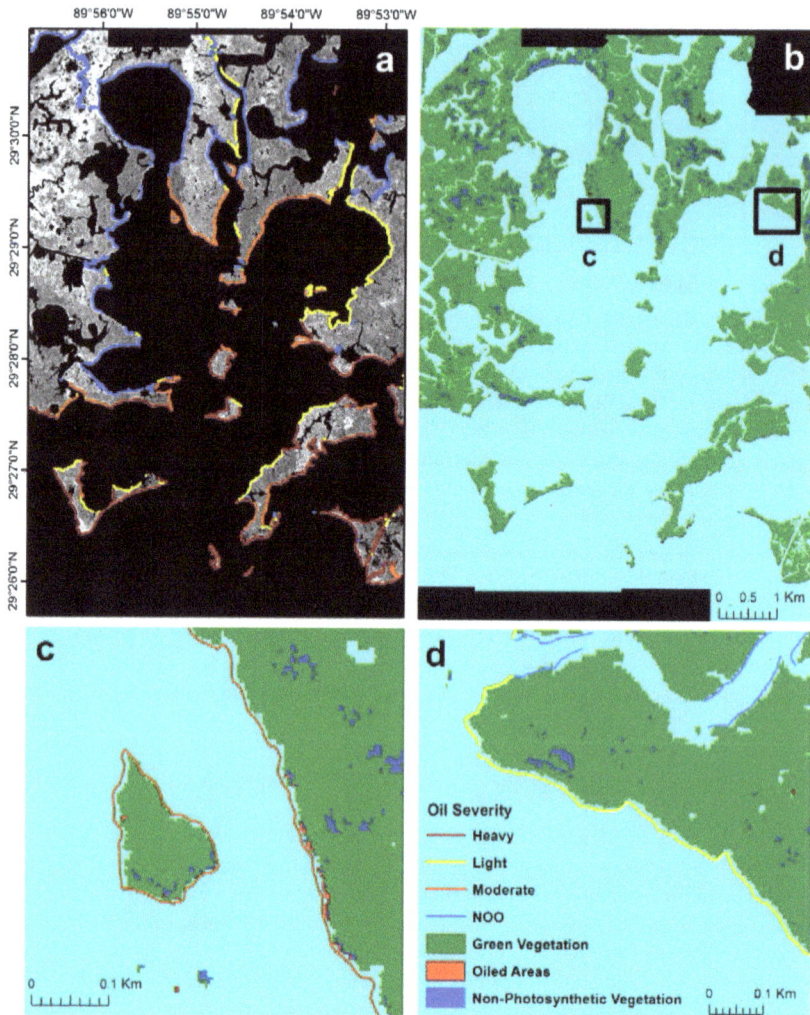

Figure 2. (**a**) SCAT oiled shoreline data overlaid on the AVIRIS flightlines acquired September 14, 2010 over Barataria Bay, (**b**) landcover classification showing location of inset c and d, (**c**) a shoreline designated as moderately oiled by SCAT, with patches of oiled pixels, and (**d**) a shoreline designated as lightly oiled by SCAT with very few oiled pixels.

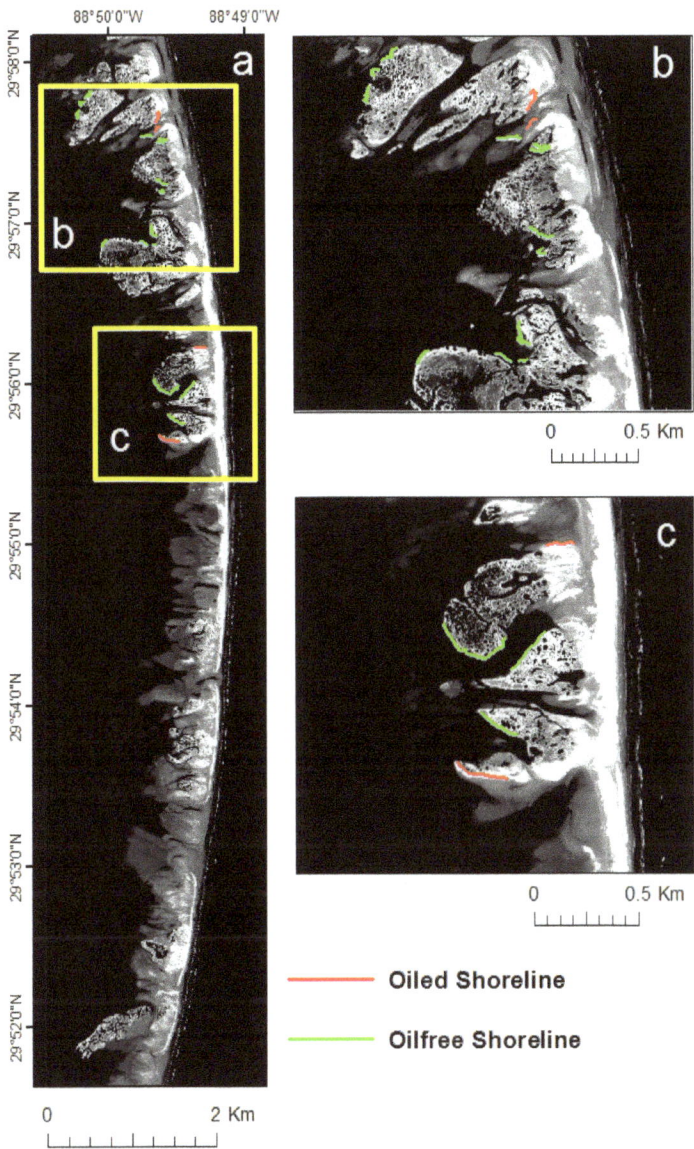

Figure 3. (**a**) SCAT oiled and oil-free shoreline data selected for analysis overlaid on the AVIRIS flightlines acquired over Chandeleur Islands on September 21, 2010 with detailed views of two sections, (**b**) and (**c**) affected by oiling.

Figure 4. (**a**) SCAT oiled and oil-free shoreline data overlaid on the AVIRIS flightlines acquired October 3, 2010 over East Bird's Foot with (**b**) detailed view of the most affected section.

Once the shorelines to be analyzed were selected, we overlaid the oiled and oil-free shoreline vectors on the index and classification images, and created a 16 m buffer inland from the shorelines. There were two reasons for this: (1) Khanna *et al.* [13] showed that in BB, oil impacts were significant within the first 16 m perpendicular to the shoreline, and (2) the bathymetry and tidal interaction in each site creates very different patterns of oiling. In BB, shallow bathymetry and the tide create different oil residence times and penetration distances, leading to differential impacts moving inland from the shore [28]. In CI, the vegetation grows on low-lying dunes on the landward side, where ocean waves are able to penetrate far across the islands. The frequent tidal overwashing could have dispersed the oil across the island, potentially causing similar oil impacts irrespective of the distance of the vegetation from the shoreline [27]. In EBF, the hydrology has been greatly modified by human activity, making it the most fragmented and developed area of all three sites. Given these differences, we identified an impact buffer for each site. The 16 m buffer translates to a 2-pixel buffer for images with 7.7 m pixels and a 5-pixel buffer for the images with 3.4 to 3.5 m pixels.

2.4.2. Comparison of Oil Impact across Sites

For the analysis of oil impact, we focused on seven spectral indices (Table 2) from different wavelength regions of the solar spectrum, ranging from the visible through the shortwave-infrared that track changes in chlorophyll content and/or leaf area of the plant (Normalized Difference Vegetation Index: NDVI and modified NDVI: mNDVI), changes in plant condition: green, senescent, dead (Angle at NIR: ANIR and Angle at Red: ARed), and changes in plant water content (Normalized Difference Infrared Index: NDII, Water Absorption at 980 nm: ADW1, and Water Absorption at 1240 nm: ADW2) (Table 2). We tested these multiple indices against the predictor variable of oiling. No single index is optimal for all conditions; plant stress can be expressed as pigment loss, water loss, leaf area loss, or all three. Consistent responses across these diverse indices indicates the analysis is more robust. Many of these indices are correlated (e.g., ANIR and ARed have two bands in common); however, since they are response variables, there is no requirement that they should be statistically independent. The differences between them highlight how different biophysiological characteristics of the wetland plant communities respond to oil-induced stress.

The red edge inflection point (REIP) [9] has been useful in detecting stress and senescence and has been used to detect stress from oil [8], but was not used in this study. Previous work in the response to the BP-DWH oil spill in Barataria Bay showed much of the intertidal vegetation was coated with oil and was dead and partially removed by tidal action at the time the imagery was acquired [13]. Thus many pixels affected by oil no longer had green foliage and the spectral response lacked the red edge feature.

We tested for significant differences between index values extracted from oiled and oil-free buffer zones at all three sites using the Mann-Whitney Test—a non-parametric test [45] (Table 3). We could not use a simple *t*-test because our index histograms were either bi-modal or skewed and violated the assumption of normality. We believe both these distributions were due to the mix of landcover types in our data. For many indices (e.g., ANIR, ARed, ADW, mNDVI), soil-NPV and green vegetation had distinct peaks resulting in a bi-modal distribution while some indices like NDVI or NDII exhibited a skewed distribution.

We also calculated Cliff's delta, a measure of effect size for non-normal data [46–48]. Cliff's delta measures the degree of overlap between two populations. It ranges from −1 to +1, where an effect size of 1 or −1 indicates no overlap in the two populations and a value of 0 means the distributions are identical. When interpreting delta, 0.147 represents a small effect (percent of non-overlap is 14.7%), 0.33 represents a medium effect (percent of non-overlap is 33%), and 0.474 represents a large effect (percent of non-overlap is 47.4%) [48]. Unlike significance tests, effect size provides a statistic that is independent of sample size and range of index values. It allowed us to standardize the difference between two populations across many variables (e.g., multiple indices) and to quantify the magnitude of a difference in effect across sites. By using effect sizes, we were able to compare the performance of different indices and their sensitivity to differences in impact across sites.

Table 3. Median values for each index for oil and oil-free shorelines, the U statistic and *p*-values from the Mann-Whitney Test and the effect size as calculated with Cliff's delta. *n* indicates total number of pixels in each study site for all oiled shorelines and oil-free shorelines.

Site	Median		U	*p*-Value	Cliff's Delta
	Oiled	Oil-Free			
BB	*n* = 36,588	*n* = 23,319			
NDVI	0.61	0.68	366,505,856	<0.0001	0.14
mNDVI	0.30	0.37	365,362,851	<0.0001	0.14
NDII	0.20	0.24	361,584,380	<0.0001	0.15
ANIR	0.95	0.50	513,898,843	<0.0001	0.20
ARed	4.67	4.99	348,463,947	<0.0001	0.18
ADW1	219.62	270.51	353,379,597	<0.0001	0.17
ADW2	414.13	484.95	352,463,292	<0.0001	0.17
CI	*n* = 3,089	*n* = 8,460			
NDVI	0.31	0.55	4,782,283	<0.0001	0.63
mNDVI	0.11	0.28	5,436,240	<0.0001	0.58
NDII	0.10	0.36	3,023,503	<0.0001	0.77
ANIR	1.56	0.70	20,080,843	<0.0001	0.54
ARed	4.71	5.56	4,765,186	<0.0001	0.64
ADW1	249.26	425.32	7,584,217	<0.0001	0.42
ADW2	374.55	621.53	7,011,257	<0.0001	0.46
EBF	*n* = 1,620	*n* = 2,982			
NDVI	3.5	0.81	2,822,096	<0.0001	0.17
mNDVI	0.68	0.68	2,663,718	<0.0001	0.10
NDII	0.63	0.63	2,428,518	0.7600	-
ANIR	0.23	0.22	2,122,563	<0.0001	0.12
ARed	5.98	5.98	2,592,410	0.0007	0.07
ADW1	438.40	510.04	2,905,034	<0.0001	0.20
ADW2	478.29	543.92	2,839,563	<0.0001	0.18

2.4.3. Comparison of Recovery from the Oil Spill across Sites

For the analysis of recovery from oil impact, we used the classified images for both years and compared them after classification. We resized the 2010 imagery to the same resolution as the 2011 imagery and overlaid the 16 m buffer used in the oil impact analysis to perform a change detection analysis. Additionally, we included the equivalent buffer distance on the seaward side (mostly water) to take into account any new growth extending outward from the shoreline after 2010. To estimate recovery we looked at total percentage of green vegetation at a site in 2010 and 2011 and any change in area covered by vegetation.

$$PCveg_{2010} = PXveg_{2010}/PXtotal_{2010}$$

where,

$PCveg_{2010}$ = percentage of green vegetation pixels in 2010
$PXveg_{2010}$ = total number of pixels of green vegetation in 2010
$PXtotal_{2010}$ = total number of pixels analyzed

Similarly,

$$PCveg_{2011} = PXveg_{2011}/PXtotal_{2011}$$

The difference between $PCveg_{2010}$ and $PCveg_{2011}$ serves as an indicator of recovery or lack thereof. This analysis was restricted to the oiled shorelines to determine if they recovered.

We tested for significant conversion from the NPV-Soil, and water classes to green vegetation and *vice versa*. Even sub-pixel level misregistration between images from different dates can lead to overestimation of change by inflating the change from one class to another [49,50]. For example, from 2010 to 2011, some areas converted from water and NPV-Soil to green vegetation and some green vegetation became water or NPV-Soil. We tested whether this exchange between classes was more than expected through random chance using McNemar's Test. This test is used for data that is not independent and examines two proportions for significant differences [51]. It is sometimes referred to as a "within-subjects chi-squared test" and the test statistic (χ^2) is a chi-squared value compared against the critical value to find significance. This test requires a 2 × 2 contingency table, thus we combined the NPV-Soil and water classes into one class called "other." The results from this test indicate whether the amount of recovery (*i.e.*, conversion of NPV-Soil/Water in 2010 to green vegetation in 2011) was significantly different from any conversion of vegetation with green foliage in 2010 to NPV-Soil/water in 2011.

3. Results

3.1. Comparison of Oil Impact across Sites

The Chandeleur Islands had the highest average Cliff's delta effect size at 0.58, which indicates large differences in vegetation health between the oiled and oil-free shorelines. Barataria Bay had an average effect size of 0.17, which is interpreted to be a small difference between populations. East Bird's Foot had an average effect size of 0.14 which indicates the difference in vegetation condition between the oil and oil-free shorelines was negligible. Across the three sites, the effect size illustrates that CI experienced much more damage in terms of plant health than the other two sites. Table 3 summarizes the results of the Mann-Whitney test and Cliff's delta for the seven spectral indices across the three sites.

Overall, no one index or group of indices (indicators of pigment, water content, or plant condition) were best at measuring oil impact across all three sites; even within index groups (e.g., leaf water content indices), the response was not uniform. Nonetheless, the indices consistently differed between oiled and unoiled pixels and gave similar values of Cliff's delta within sites and prominent differences between sites (Figure 5).

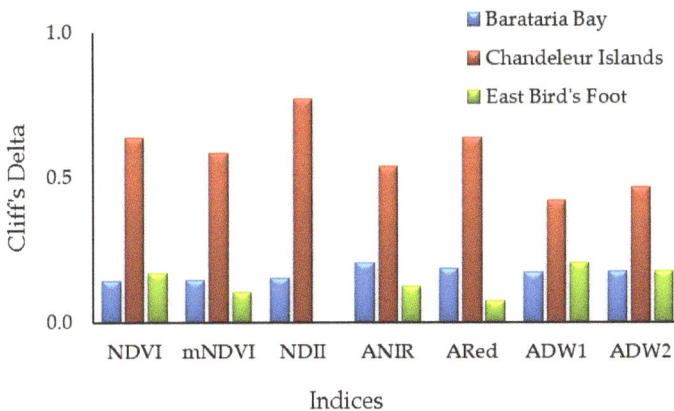

Figure 5. The difference in effect size (Cliff's delta) between the three sites for seven spectral indices.

3.2. Comparison of Recovery from Oil across Sites

For BB, we saw an increase in the percentage of green vegetation, from 34.6% to 48.1%, as the marsh recovered between 2010 to 2011 (Table 4). In CI, there was a large increase in the total percentage of green vegetation, from 30.9% in 2010 to 65.1% in 2011. In EBF, the change in green vegetation was minimal in the impacted zone (Table 4).

Table 4. Change in total classified area from 2010 to 2011 for each study site. Area of each class is in m^2 (each pixel = 59.3 m^2 for Barataria Bay and Chandeleur Islands and 11.56 m^2 for East Bird's Foot). PCveg is the overall percentage of green vegetation from 2010 to 2011 for each study site along the oiled shorelines. As the buffer zone includes both the landward side and seaward side, having roughly 50% green vegetation would indicate a good recovery on the landward side of the shoreline; greater than 50% recovery indicates water pixels becoming vegetated.

Class	Barataria Bay		Chandeleur Islands		East Bird's Foot	
	2010	2011	2010	2011	2010	2011
Green Vegetation	199,985	278,078	9,724	20,455	29,351	31,015
NPV-Soil	69,191	14,763	9,071	10,672	913	208
Water	309,375	285,719	12,629	296	33,050	32,091
PCveg	34.6%	48.1%	30.9%	65.1%	46.4%	49.0%

There was a significant increase in green vegetation in 2011 at all three sites, although the degree of increase was site dependent (Table 4). In BB, vegetation increased by 13.5% (McNemar = 740.43 and a *p*-value < 0.0001). Of the area classified as NPV-Soil in 2010, 89.0% was revegetated in 2011, and 10.2% became water. Of the area classified as NPV-Soil in 2011, 90.8% had been previously classified as green vegetation in 2010. Thus, even though there was an overall increase in green vegetation, there was still turnover between the vegetation and NPV-soil classes. An important point to note is that in BB the NPV-Soil class only represented 12.0% of the classified area in 2010 and 5.2% in 2011.

In CI, the recovery was extensive; the area of green vegetation more than doubled from 2010 to 2011 (McNemar = 123.19, *p*-value < 0.0001). The area classified as NPV-Soil increased by 5.1%, with most of this change due to 2010 water pixels becoming NPV-Soil in 2011.The area classified as water decreased 98% from 2010 to 2011, with the majority pixels transitioning to green vegetation (Table 4). Much of this increase can be attributed to lower tide levels in the 2011 imagery, which made the vegetation easier to detect and explains why NPV-Soil replaced water. The green vegetation on these exposed sandbars may have been present in 2010, but because of the tides, it was not visible in the imagery.

We further investigated this "new" vegetation by comparing it to reference spectra of two dominant wetland species and performing a spectral mixture analysis to determine the proportion of soil and water in these pixels. The spectra of the "new" vegetation growth did not match the spectral signatures of either *S. alterniflora* or *A. germinans*. Our spectral mixture analysis found these pixels to be approximately 40% to 70% vegetation, with the rest being soil. These mixed vegetation-soil spectra made it impossible to identify the new vegetation at the species level.

In EBF, there was less change compared to the other two sites (McNemar = 19.81, *p*-value < 0.0001). Green vegetation increased by 2.6% from 2010 to 2011, and there was a steep decline in the area classified as NPV-Soil class, the majority of which was classified as green vegetation in 2011 (Table 4).

4. Discussion

The BP-DWH oil spill affected many sensitive ecosystems along the 1773 km of shoreline where oil was documented [33]. The AVIRIS data flown by NASA in response to this environmental disaster provided a rare opportunity to study the impact on these diverse ecosystems using hyperspectral imagery. Our study revealed the differential response of wetland plant communities to the apparently same level of oiling.

4.1. Comparison of Oil Impact across Sites

The tides do not inundate the entire upper marsh in BB except during the highest winter tides and in catastrophic events like hurricanes. BB's topography helped restrict oil pentration into the marsh to the highest tide level. Most of the oil was located within the first 3–4 pixels (10–15 m) from the shore which is mainly saltmarsh meadow dominated by two species, *S. alterniflora* and *J. roemerianus*. Both species are known for resilience to oil contamination [52–54], meaning that they recover from short-term exposures to oil, although long-term exposure may result in death [52]. DeLaune *et al.* [52] and Kirby and Grosselink [55], both showed that application of crude oil to *S. alterniflora* did not significantly impact plant biomass, stem density or new growth. However, when oil coats the leaf surface, productivity drops due to lack of gas exchange [53,54]. Similarly, for *J. roemerianus*, photosynthetic activity decreased when the plants were partially coated with oil and ceased when inundated completely. Photosynthetic activity started recovering after four weeks [54].

The first 7 m along the shore in BB were much more likely to contain vegetation with oil-covered stems and leaves, but further inland, the oil mainly coated the soil (Khanna, personal observation, September 2010). Additionally, Khanna *et al.* [13] showed that oil impact decreased the further the pixel was from the shore. To standardize our analysis across the three sites, we quantified oil impact in a single 16 m band along the shoreline. Thus, it is likely that the "small effect" in BB is due to the zone of severe impact being restricted to the first few meters adjacent to the shoreline and being combined with a much weaker impact further inland. Additionally, BB had the largest amount of shoreline for analysis of the three study sites. A larger shoreline for analysis is bound to encompass a greater range of impact, reducing the magnitude of the difference between the oiled and oil-free shorelines. Lastly, because of non-normal distributions for all the indices, we used a non-parametric test for significance. Non-parametric tests usually have less statistical power and thus can dampen differences between populations.

In CI, the dominant species were *A. germinans* and *S. alterniflora*, but their composition varied greatly across our study area. Some shorelines were co-dominated by *A. germinans* and *S. alterniflora* while other shorelines were mainly *S. alterniflora* interspersed with *A. germinans* (Figure 6). The resilience of *S. alterniflora* to oil contamination demonstrates a stark contrast to *A. germinans*, which, in general, is highly sensitive. Oil can disrupt ion transport in mangroves which is necessary for salt exclusion, an important mechanism for mangroves living in high salinity conditions. Other sub-lethal effects include branching of pneumatophores, decreased canopy cover, increased rate of mutation, inhibited respiration, and increased sensitivity to other stresses [56]. The mangroves grow in both high and low marshes in CI. The high marshes exist above the high, neap-tide line and are inundated only at especially high tides and storm surges, but the low marshes are inundated daily at high tide [57]. Thus mangroves can be impacted by oil irrespective of the distance from the shoreline if oil reaches their roots and pneumatophores. This is in contrast to BB where the oil injured or killed plants in the first few meters of the shoreline but had a weaker impact further inland.

Our analysis showed that EBF had the weakest impact of all three study sites. In EBF, the oiled sites were located in intermediate and freshwater marshes and were dominated almost entirely by *P. australis*. Previous reasearch has found that oiling of *P. australis* can decrease biomass, stem density, photosynthetic rates, and can cause the death of emerging buds, hindering reproduction [58–60]. However, Judy *et al.* [23] investigated the impact of the BP-DWH oil on *P. australis* in a greenhouse study. Their study found that applying the BP-DWH oil to the plant shoots had no significant negative impacts on plant growth and instead induced a vegetative stress response where *P. australis* produced side shoots. In contrast, when oil was applied directly to the soil, it produced a signicant reduction in plant growth. In the case of our study, field observations by Kokaly *et al.* [34] indicated that in EBF, oil came in with the water but did not settle on the soil. Kokaly *et al.* [34] further observed that the oil sometimes coated the leaves and stems of *P. australis* but did not damage the plant canopy. This differs from BB where the oil consistently covered the plants completely at the marsh edge forcing the vegetation to flatten under the weight of the oil [34]. We also hypothesize that because EBF continued to receive fresh water from the Mississippi while the oil was coming ashore, it decreased or limited the residence time of the oil in the marsh.

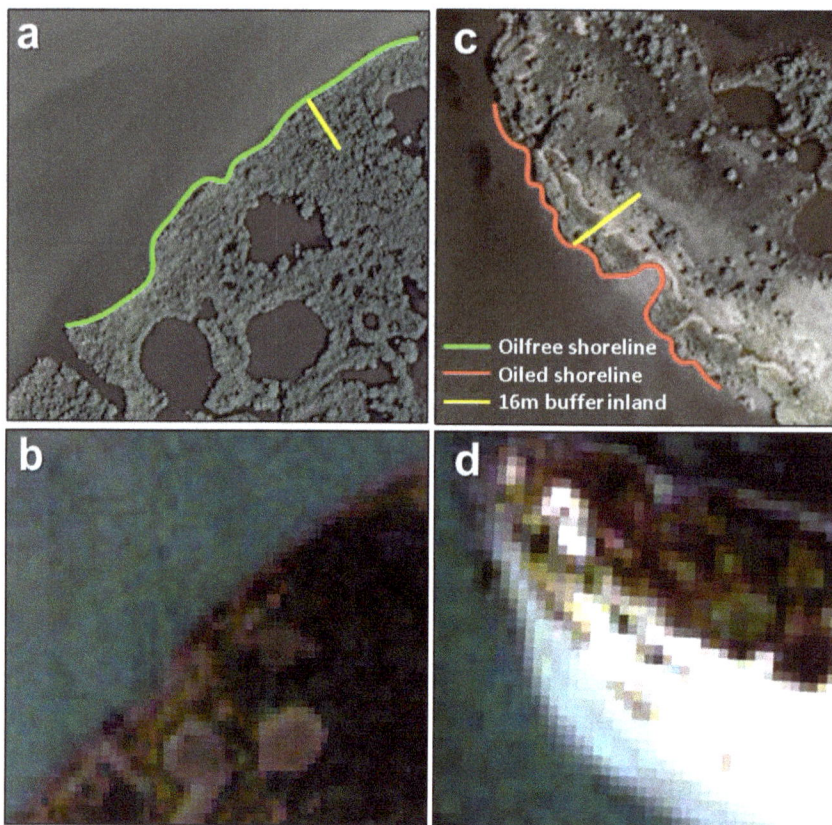

Figure 6. (a) SCAT shoreline data overlaid on 2010 GIS Aerometric 0.3 m imagery of a section of Chandeleur Islands with a yellow line marking the extent of the 16 meter buffer used in the analyses, (b) AVIRIS true color image showing the same extent of (a) at 7.7 m resolution, (c) SCAT shoreline data overlaid on 0.3m imagery of a section of Chandeleur Islands, (d) AVIRIS true color image showing the extent of (c) at 7.7 m resolution.

Our results indicate that CI was the most severely impacted of the three sites. However, CI also had the smallest subset of shoreline for analysis (Table 1). This means that the total length of shoreline affected in CI was small but the affected shore showed greater impact compared to the other two sites.

4.2. Comparison of Recovery from Oil across Sites

In BB, we saw a strong but incomplete recovery from 2010 to 2011. Several studies have shown that the dominant species in salt marshes are resilent to oiling in moderate to low doses and have good regrowth after canopy mortality under heavy oiling [7,24,53,61,62]. The lack of a full recovery in BB may be a consequence of looking at the mean response which is a combination of the strong impact that occurred in the first two pixels and the lower impact in the interior of the marsh.

CI exhibited the largest change of the three sites. Green vegetation more than doubled and a large majority of this newly vegetated area is a result of areas of water being converted to green vegetation.

Unlike the other two sites, these islands are constantly changing and being reshaped by wave action and as the underlying sand is reworked and deposited, new areas of land emerge as colonization sites [63]. Another contributing factor to the large change in green vegetation between years is the differing tide levels in the imagery (0.66 m in 2010 and 0.25 m in 2011). This tidal difference may explain some of the new vegetation found in 2011; however, the increase in green vegetation along oil-free shorelines was less than that along the oiled shorelines. Thus, tidal height difference does not explain all of the new vegetation in 2011.

The two dominant species on Chandeleur Islands, *S. alterniflora* and *A. germinans*, have different responses to oiling. Previous studies have shown that crude oil is highly detrimental to mangrove reproduction and easily kills mangrove seedlings [64]. Thus, one would not expect mangroves to show a strong recovery based on seedling establishment one year after oiling. As discussed in Section 3.2, spectral unmixing indicated a combination of soil plus vegetation, but the vegetation endmember signature could not be identified clearly. The tidal difference between the image dates may overestimate recovery, but new vegetation growth at CI is more likely from *S. alterniflora* rather than new mangroves.

In EBF, we saw a slight increase of green vegetation in 2011. This minimal response is consistent with the low impact observed in the vegetation indices. Having a constant supply of fresh water from the Mississippi River likely aided recovery in addition to lessening the original impact of oil by more rapid removal from the marsh.

5. Conclusions

The British Petroleum Deepwater Horizon oil spill affected a wide expanse of coastal ecosystems. Our examination of three ecologically diverse sites found different levels of vulnerability and resilience to oil. While Barataria Bay was the most extensively oiled site, it was more resilient to the deleterious effects of oil contamination compared to Chandeleur Islands which quantitatively had the least amount of heavily oiled shoreline. The impact on each site relied on the interplay between its topography, hydrology, and plant response. Hyperspectral imagery allowed us to track the changes in plant physiology across a spatially extensive area and over multiple years. This research has implications for future response to oil spills by highlighting the fact that areas subject to the most oil contamination may not be those most vulnerable to oil. Hence management decisions regarding clean-up and remediation need to take into account both the extent/severity of contamination and the sensitivity of ecosystems to that contamination. The Mississippi Deltaic Plain is still recovering from the effects of the 2010 oil spill and more studies are needed to look at long-term impacts across the many ecosystems that were affected. AVIRIS imagery has already been acquired over Barataria Bay in both 2012 and 2015 and analysis of this imagery currently underway will shed critical light on the long-term resilience of an ecosystem after an environmental disaster. Remotely sensed imagery is a powerful tool that can help cover the wide expanse of environmental disasters and document plant recovery in the following years.

Acknowledgments: This work was funded by a NASA Terrestrial Ecology grant (#NNX12AK58G). We thank Diane Wickland, program manager for Terrestrial Ecology at NASA, for providing access to the AVIRIS database and to the AVIRIS team for preprocessing support. We thank George Scheer for systems support. We also thank Maria J. Santos for offering advice and helping revise our text.

Author Contributions: K. Shapiro researched, conceived, and performed the analysis, and wrote the paper. S. Khanna contributed to all aspects of the analysis, provided programming expertise, edited the paper, and advised on figure and table design. This research was conducted under the guidance of S. Ustin in her lab and she also contributed to the editing and organization of the paper.

Conflicts of Interest: The authors declare no conflict of interest. The funding agency had no role in the design of the study; in the collection, analyses, or interpretation of data; in the writing of the manuscript; and in the decision to publish the results.

J. Mar. Sci. Eng. **2016**, *4*, 33

Abbreviations

The following abbreviations are used in this manuscript:

ADW1   Absorption Depth of Water at 980 nm
ADW2   Absorption Depth of Water at 1240 nm
ANIR   Angle formed at NIR
ANOVA   ANalysis Of VAriance
ARed   Angle formed at Red
AVIRIS Airborne Visible/Infrared Imaging Spectrometer
BB   Barataria Bay
BP-DWH British Petroleum—DeepWater Horizon (oil spill)
CI   Chandeleur Islands
EBF   East Bird's Foot
ERMA   Environmental Response Management Application
MDP   Mississippi Deltaic Plain
mNDVI   modified Normalized Difference Vegetation Index
NASA   National Aeronautics and Space Administration
NDII   Normalized Difference Infrared Index
NDVI   Normalized Difference Vegetation Index
NIR   Near InfraRed
NOAA   National Oceanic and Atmospheric Administration
NPV   Non-photosynthetic vegetation
REIP   Red edge Inflexion Point
SCAT   Shoreline Cleanup Assessment Technique

References

1. Wiens, J.A. Review of an ecosystem services approach to assessing the impacts of the Deepwater Horizon Oil Spill in the Gulf of Mexico. *Fisheries* **2015**, *40*, 86–86. [CrossRef]
2. Ko, J.Y.; Day, J.W. A review of ecological impacts of oil and gas development on coastal ecosystems in the Mississippi Delta. *Ocean Coast. Manag.* **2004**, *47*, 597–623. [CrossRef]
3. Gosselink, J.G.; Pendleton, E.C. *The Ecology of Delta Marshes of Coastal Louisiana: A Community Profile*; U.S. Fish and Wildlife Service: Washington, D.C., USA, 1984; p. 156.
4. Couvillion, B.R.; Barras, J.A.; Steyer, G.D.; Sleavin, W.; Fischer, M.; Beck, H.; Trahan, N.; Griffin, B.; Heckman, D. *Land Area Change in Coastal Louisiana (1932 to 2010)*; USGS National Wetlands Research Center: Lafayette, LA, USA, 2011.
5. Silliman, B.R.; van de Koppel, J.; McCoy, M.W.; Diller, J.; Kasozi, G.N.; Earl, K.; Adams, P.N.; Zimmerman, A.R. Degradation and resilience in Louisiana salt marshes after the BP-Deepwater Horizon oil spill. *Proc. Natl. Acad. Sci. USA* **2012**, *109*, 11234–11239. [CrossRef] [PubMed]
6. Hester, M.W.; Mendelssohn, I.A. Long-term recovery of a Louisiana brackish marsh plant community from oil-spill impact: Vegetation response and mitigating effects of marsh surface elevation. *Mar. Environ. Res.* **2000**, *49*, 233–254. [CrossRef]
7. Pezeshki, S.R.; Hester, M.W.; Lin, Q.; Nyman, J.A. The effects of oil spill and clean-up on dominant US Gulf coast marsh macrophytes: A review. *Environ. Pollut.* **2000**, *108*, 129–139. [CrossRef]
8. Li, L.; Ustin, S.L.; Lay, M. Application of AVIRIS data in detection of oil-induced vegetation stress and cover change at Jornada, New Mexico. *Remote Sens. Environ.* **2005**, *94*, 1–16. [CrossRef]
9. Horler, D.N.H.; Dockray, M.; Barber, J. The red edge of plant leaf reflectance. *Int. J. Remote Sens.* **1983**, *4*, 273–288. [CrossRef]
10. Gitelson, A.; Merzlyak, M.N. Spectral reflectance changes associated with autumn senescence of *Aesculus hippocastanum* L. and *Acer platanoides* L. Leaves. Spectral features and relation to chlorophyll estimation. *J. Plant Physiol.* **1994**, *143*, 286–292. [CrossRef]
11. Tucker, C.J. Red and photographic infrared linear combinations for monitoring vegetation. *Remote Sens. Environ.* **1979**, *8*, 127–150. [CrossRef]

12. Hunt, E.R.; Rock, B.N. Detection of changes in leaf water content using near-infrared and middle-infrared reflectances. *Remote Sens. Environ.* **1989**, *30*, 43–54.
13. Khanna, S.; Santos, M.J.; Ustin, D.S.L.; Koltunov, A.; Kokaly, R.F.; Roberts, D.A. Detection of salt marsh vegetation stress after the Deepwater Horizon BP oil spill along the shoreline of gulf of Mexico using AVIRIS data. *PloS ONE* **2013**, *8*, e78989. [CrossRef] [PubMed]
14. Khanna, S.; Palacios-Orueta, A.; Whiting, M.L.; Ustin, S.L.; Riano, D.; Litago, J. Development of angle indexes for soil moisture estimation, dry matter detection and land-cover discrimination. *Remote Sens. Environ.* **2007**, *109*, 154–165. [CrossRef]
15. Dibner, P.C. *Response of A Salt Marsh to Oil Spill and Cleanup: Biotic and Erosional Effects in the Hackensack Meadowlands, New Jersey. Final report, May 1976–December 1977*; URS Research Co.: San Mateo, CA, USA, 1978.
16. Long, B.F.; Vandermeulen, J.H. Geomorphological impact of cleanup of an oiled salt marsh (Ile Grande, France). In Proceedings of the International Oil Spill Conference, San Antonio, Texas, USA, 28 February–3 March 1983; pp. 501–505.
17. Mishra, D.R.; Cho, H.J.; Ghosh, S.; Fox, A.; Downs, C.; Merani, P.B.T.; Kirui, P.; Jackson, N.; Mishra, S. Post-spill state of the marsh: Remote estimation of the ecological impact of the Gulf of Mexico oil spill on Louisiana Salt Marshes. *Remote Sens. Environ.* **2012**, *118*, 176–185. [CrossRef]
18. Rosso, P.H.; Pushnik, J.C.; Lay, M.; Ustin, S.L. Reflectance properties and physiological responses of *Salicornia virginica* to heavy metal and petroleum contamination. *Environ. Pollut.* **2005**, *137*, 241–252. [CrossRef] [PubMed]
19. Van der Meer, F.; van Dijk, P.; van der Werff, H.; Yang, H. Remote sensing and petroleum seepage: A review and case study. *Terra Nova* **2002**, *14*, 1–17. [CrossRef]
20. Yang, H.; Zhang, J.; van der Meer, F.; Kroonenberg, S.B. Geochemistry and field spectrometry for detecting hydrocarbon microseepage. *Terra Nova* **1998**, *10*, 231–235. [CrossRef]
21. Bammel, B.H.; Birnie, R.W. *Spectral Reflectance Response of Big Sagebrush to Hydrocarbon-Induced Stress in the Bighorn Basin, Wyoming*; American Society for Photogrammetry and Remote Sensing: Bethesda, MD, USA, 1994; Volume 60.
22. Khanna, S.; Santos, M.J.; Ustin, S.L.; Haverkamp, P.J. An integrated approach to a biophysiologically based classification of floating aquatic macrophytes. *Int. J. Remote Sens.* **2011**, *32*, 1067–1094. [CrossRef]
23. Judy, C.R.; Graham, S.A.; Lin, Q.; Hou, A.; Mendelssohn, I.A. Impacts of Macondo oil from Deepwater Horizon spill on the growth response of the common reed *Phragmites australis*: A mesocosm study. *Mar. Pollut. Bull.* **2014**, *79*, 69–76. [CrossRef] [PubMed]
24. Lin, Q.; Mendelssohn, I.A. Impacts and recovery of the Deepwater Horizon Oil Spill on vegetation structure and function of coastal salt marshes in the northern gulf of Mexico. *Environ. Sci. Technol.* **2012**, *46*, 3737–3743. [CrossRef] [PubMed]
25. Wu, W.; Biber, P.D.; Peterson, M.S.; Gong, C. Modeling photosynthesis of *Spartina alterniflora* (smooth cordgrass) impacted by the Deepwater Horizon oil spill using Bayesian inference. *Environ. Res. Lett.* **2012**, *7*, 045302. [CrossRef]
26. Visser, J.; Sasser, C.; Chabreck, R.; Linscombe, R.G. Marsh vegetation types of the Mississippi River Deltaic Plain. *Estuaries* **1998**, *21*, 818–828. [CrossRef]
27. Hymel, M. *Monitoring Plan for Chandeleur Islands Marsh Restoration*; LDNR/Coastal Restoration and Management: Baton Rouge, LA, USA, 2001; p. 12.
28. Day, J.; Britsch, L.; Hawes, S.; Shaffer, G.; Reed, D.; Cahoon, D. Pattern and process of land loss in the Mississippi Delta: A Spatial and temporal analysis of wetland habitat change. *Estuaries* **2000**, *23*, 425–438. [CrossRef]
29. Wilson, C.A.; Allison, M.A. An equilibrium profile model for retreating marsh shorelines in southeast Louisiana. *Estuar. Coast. Shelf Sci.* **2008**, *80*, 483–494. [CrossRef]
30. NOAA. Deepwater Horizon Data Integration Visualization Exploration and Reporting Application. Available online: http://dwhdiver.orr.noaa.gov (accessed on 31 January 2016).
31. CWPPRA. The Mississippi River Delta Basin. Available online: http://lacoast.gov/new/About/Basin_data/mr/ (accessed on 31 January 2016).
32. USGS. Coastwide Reference Monitoring System. Available online: http://lacoast.gov/crms2/ (accessed on 31 January 2016).

33. Michel, J.; Owens, E.H.; Zengel, S.; Graham, A.; Nixon, Z.; Allard, T.; Holton, W.; Reimer, P.D.; Lamarche, A.; White, M.; *et al.* Extent and degree of shoreline oiling: *Deepwater Horizon* oil spill, Gulf of Mexico, USA. *PLoS ONE* **2013**, *8*, e65087. [CrossRef] [PubMed]

34. Kokaly, R.F.; Heckman, D.; Holloway, J.; Piazza, S.C.; Couvillion, B.R.; Steyer, G.D.; Mills, C.T.; Hoefen, T.M. *Shoreline Surveys of Oil-Impacted Marsh in Southern Louisiana, July to August 2010: Open-File Report 2011-1022*; USGS Crustal Geophysics and Geochemistry Science Center: Denver, CO, USA, 2011.

35. Cretini, K.F.; Visser, J.M.; Krauss, K.W.; Steyer, G.D. *CRMS Vegetation Analytical Team Framework: Methods for Collection, Development, and Use of Vegetation Response Variables*; USGS National Wetlands Research Center: Lafayette, LA, USA, 2011; p. 60.

36. Palacios-Orueta, A.; Khanna, S.; Litago, J.; Whiting, M.L.; Ustin, S.L. Assessment of NDVI and NDWI spectral indices using MODIS time series analysis and development of a new spectral index based on MODIS shortwave infrared bands. In Proceedings of the 1st International Conference of Remote Sensing and Geoinformation Processing, Trier, Germany, 7–9 September 2005.

37. Friedl, M.A.; Brodley, C.E. Decision tree classification of land cover from remotely sensed data. *Remote Sens. Environ.* **1997**, *61*, 399–409. [CrossRef]

38. Hickman, J.C. *The Jepson Manual: Higher Plants of California*; University of California Press: Berkeley, CA, USA, 1993.

39. Biello, M.; Rust, D.; Watkins, T. Oil laps barrier islands; BP grilled about oil spill at Capitol. Available online: http://www.cnn.com (accessed on 4 May 2010).

40. Guillot, C. Oil spill hits gulf coast habitats. Available online: http://news.nationalgeographic.com (accessed on 30 April 2010).

41. Strassmann, M. The fight over keeping oil out of Barataria Bay. Available online: http://www.cbsnews.com (accessed on 7 July 2010).

42. Ustin, S.L.; Roberts, D.A.; Khanna, S.; Shapiro, K.; Beland, M.; Peterson, S.; Roth, K. *Coastal Wetland and Near Shore Ecosystem Impacts from the Gulf of Mexico Deepwater Horizon BP Oil Spill Monitored by NASA's AVIRIS and MASTER Imagers*; NASA: Davis, CA, USA, 2014.

43. Kokaly, R.F.; Couvillion, B.R.; Holloway, J.M.; Roberts, D.A.; Ustin, S.L.; Peterson, S.H.; Khanna, S.; Piazza, S.C. Spectroscopic remote sensing of the distribution and persistence of oil from the Deepwater Horizon spill in Barataria Bay marshes. *Remote Sens. Environ.* **2013**, *129*, 210–230. [CrossRef]

44. Peterson, S.H.; Roberts, D.A.; Beland, M.; Kokaly, R.F.; Ustin, S.L. Oil detection in the coastal marshes of Louisiana using MESMA applied to band subsets of AVIRIS data. *Remote Sens. Environ.* **2015**, *159*, 222–231. [CrossRef]

45. Mann, H.B.; Whitney, D.R. On a test of whether one of two random variables is stochastically larger than the other. *Ann. Math. Stat.* **1947**, *18*, 50–60. [CrossRef]

46. Macbeth, G.; Razumiejczyk, E.; Ledesma, R.D. Cliff's Delta Calculator: A non-parametric effect size program for two groups of observations. *Univ. Psychol.* **2011**, *10*, 545–555.

47. Cliff, N. Dominance statistics: Ordinal analyses to answer ordinal questions. *Psychol. Bull.* **1993**, *114*, 494–509. [CrossRef]

48. Romano, J.; Kromrey, J.D.; Coraggio, J.; Skowronek, J.; Devine, L. Exploring methods for evaluating group differences on the NSSE and other surveys: Are the t-test and Cohen's d indices the most appropriate choices? In Proceedings of the Annual Meeting of the Southern Association for Institutional Research, Arlington, VA, USA, 14–17 October 2006.

49. Dai, X.L.; Khorram, S. The effects of image misregistration on the accuracy of remotely sensed change detection. *ITGRS* **1998**, *36*, 1566–1577.

50. Khorram, S.; Biging, G.S.; Chrisman, N.R.; Colby, D.R.; Congalton, R.G.; Dobson, J.E.; Ferguson, R.L.; Goodchild, M.R.; Jensen, J.R.; Mace, T.H. *Accuracy Assessment of Remote Sensing Derived Change Detection*; American Society of Photogrammetry and Remote Sensing (ASPRS): Bethesda, MD, USA, 1998.

51. McNemar, Q. Note on the sampling error of the difference between correlated proportions or percentages. *Psychometrika* **1947**, *12*, 153–157. [CrossRef] [PubMed]

52. DeLaune, R.D.; Patrick, W.H., Jr.; Buresh, R.J. Effect of crude oil on a Louisiana *Spartina alterniflora* salt marsh. *Environ. Pollut.* **1979**, *20*, 21–31. [CrossRef]

53. DeLaune, R.D.; Pezeshki, S.R.; Jugsujinda, A.; Lindau, C.W. Sensitivity of US Gulf of Mexico coastal marsh vegetation to crude oil: Comparison of greenhouse and field responses. *Aquat. Ecol.* **2003**, *37*, 351–360. [CrossRef]

54. Pezeshki, S.R.; DeLaune, R.D. Effect of crude oil on gas exchange functions of *Juncus roemerianus* and *Spartina alterniflora*. *Water Air Soil Pollut.* **1993**, *68*, 461–468. [CrossRef]

55. Kirby, C.J.; Gosselink, J.G. Primary Production in a Louisiana Gulf Coast *Spartina Alterniflora* Marsh. *Ecology* **1976**, *57*, 1052–1059. [CrossRef]

56. Hoff, R.; Hensel, P.; Proffitt, E.C.; Delgado, P.; Shigenaka, G.; Yender, R.; Mearns, A.J. *Oil Spills in Mangroves: Planning & Response Considerations*; U.S. Department of Commerce, NOAA: Seattle, WA, USA, 2010; p. 72.

57. Hester, M.W.; Spalding, E.A.; Franze, C.D. Biological resources of the Louisiana Coast: Part 1. An overview of coastal plant communities of the Louisiana gulf shoreline. *J. Coast. Res.* **2005**, 134–145. Available online: http://www.jstor.org/stable/25737053 (accessed on 22 April 2016).

58. Dowty, R.A.; Shaffer, G.P.; Hester, M.W.; Childers, G.W.; Campo, F.M.; Greene, M.C. Phytoremediation of small-scale oil spills in fresh marsh environments: A mesocosm simulation. *Mar. Environ. Res.* **2001**, *52*, 195–211. [CrossRef]

59. Lin, Q.; Mendelssohn, I.A.; Hester, M.W.; Webb, E.C.; Henry, J.; Charles, B. Effect of oil cleanup methods on ecological recovery and oil degradation of *Phragmites* marshes. *Int. Oil Spill Conf. Proc.* **1999**, *1999*, 511–517. [CrossRef]

60. Armstrong, J.; Keep, R.; Armstrong, W. Effects of oil on internal gas transport, radial oxygen loss, gas films and bud growth in *Phragmites australis*. *Ann. Bot.* **2009**, *103*, 333–340. [CrossRef]

61. DeLaune, R.D.; Wright, A.L. Projected impact of Deepwater Horizon oil spill on U.S. gulf coast wetlands. *Soil Sci. Soc. Am. J.* **2011**, *75*, 1602–1612. [CrossRef]

62. Smith, C.J.; Delaune, R.D.; Patrick, W.H.; Fleeger, J.W. Impact of dispersed and undispersed oil entering a gulf coast salt marsh. *Environ. Toxicol. Chem.* **1984**, *3*, 609–616. [CrossRef]

63. Hayden, B.P.; Santos, M.C.F.V.; Shao, G.; Kochel, R.C. Geomorphological controls on coastal vegetation at the Virginia Coast Reserve. *Geomo* **1995**, *13*, 283–300. [CrossRef]

64. Grant, D.L.; Clarke, P.J.; Allaway, W.G. The response of grey mangrove (*Avicennia marina* (Forsk.) Vierh.) seedlings to spills of crude oil. *J. Exp. Mar. Biol. Ecol.* **1993**, *171*, 273–295. [CrossRef]

Journal of
Marine Science and Engineering

MDPI

Article

Gene Transcript Profiling in Sea Otters Post-*Exxon Valdez* Oil Spill: A Tool for Marine Ecosystem Health Assessment

Lizabeth Bowen [1,*], A. Keith Miles [1], Brenda Ballachey [2], Shannon Waters [1] and James Bodkin [2]

[1]   U.S. Geological Survey, Western Ecological Research Center, Davis Field Station, University of California, Davis, CA 95616, USA; keith_miles@usgs.gov (A.K.M.); swaters@usgs.gov (S.W.)
[2]   U.S. Geological Survey, Alaska Science Center, Anchorage, AK 99508, USA; bballachey@shaw.ca (B.B.); jbodkin@usgs.gov (J.B.)
*   Correspondence: lbowen@usgs.gov; Tel.: +1-530-752-5365; Fax: +1-916-278-9475

Academic Editor: Merv Fingas
Received: 8 January 2016; Accepted: 16 May 2016; Published: 1 June 2016

Abstract: Using a panel of genes stimulated by oil exposure in a laboratory study, we evaluated gene transcription in blood leukocytes sampled from sea otters captured from 2006–2012 in western Prince William Sound (WPWS), Alaska, 17–23 years after the 1989 *Exxon Valdez* oil spill (EVOS). We compared WPWS sea otters to reference populations (not affected by the EVOS) from the Alaska Peninsula (2009), Katmai National Park and Preserve (2009), Clam Lagoon at Adak Island (2012), Kodiak Island (2005) and captive sea otters in aquaria. Statistically, sea otter gene transcript profiles separated into three distinct clusters: Cluster 1, Kodiak and WPWS 2006–2008 (higher relative transcription); Cluster 2, Clam Lagoon and WPWS 2010–2012 (lower relative transcription); and Cluster 3, Alaska Peninsula, Katmai and captive sea otters (intermediate relative transcription). The lower transcription of the aryl hydrocarbon receptor (AHR), an established biomarker for hydrocarbon exposure, in WPWS 2010–2012 compared to earlier samples from WPWS is consistent with declining hydrocarbon exposure, but the pattern of overall low levels of transcription seen in WPWS 2010–2012 could be related to other factors, such as food limitation, pathogens or injury, and may indicate an inability to mount effective responses to stressors. Decreased transcriptional response across the entire gene panel precludes the evaluation of whether or not individual sea otters show signs of exposure to lingering oil. However, related studies on sea otter demographics indicate that by 2012, the sea otter population in WPWS had recovered, which indicates diminishing oil exposure.

Keywords: gene transcription; *Exxon Valdez* oil spill; sea otter; *Enhydra lutris*; oil exposure; Prince William Sound; recovery

1. Introduction

The effects of the 1989 *Exxon Valdez* oil spill (EVOS) on nearshore marine vertebrates in Prince William Sound, Alaska, including the sea otter (*Enhydra lutris*), have continued for more than two decades [1–9]. A series of long-term studies demonstrated a lack of recovery of sea otters through at least 2009 [1,4,6–9], based on reduced rates of survival and exposure to residual oil in western Prince William Sound (WPWS), although the importance of continuing exposure as a factor constraining sea otter recovery has been debated [10–12]. To evaluate population health and recovery of sea otters and other species potentially affected by the spill, the *Exxon Valdez* Oil Spill Trustee Council established physiologic (based on biomarkers indicating exposure to aromatic hydrocarbons) and demographic (based on a return to expected abundance or reproduction/survival rates) criteria.

We used molecular gene transcription to examine the physiological status of sea otters in oiled areas of WPWS, the geographic region most severely affected by the oil spill. Exposure to petroleum hydrocarbons has the potential to cause not only catastrophic short-term effects, but importantly, often overlooked, long-term damage to individuals, populations or even ecosystems [1,13]. The extent and duration of long-term effects are difficult to assess, as pathophysiological changes within an individual may be significant yet subtle and, consequently, undetectable using classical wildlife diagnostic methods. Alterations in the levels of gene transcription can provide the earliest observable signs of health impairment, discernable prior to clinical manifestation [14–16]. The utility of the methodology used in our study relies on the assumption that oil-induced pathology in sea otters is accompanied by predictable and specific changes in gene transcription.

In 2008, we sampled sea otters in previously oiled and unoiled areas of WPWS and compared these to samples from reference (*i.e.*, deemed clinically normal) sea otters from the Alaska Peninsula and captive, healthy sea otters from aquaria [6]. We concluded that sea otters in oiled areas had gene transcription patterns consistent with chronic, low-grade exposure to organic compounds. In 2010 and 2012, we resampled sea otters in the same areas of WPWS to evaluate whether gene transcription patterns observed in 2008 persisted. To provide a broader geographic and temporal interpretation for the analysis of WPWS samples collected in 2010 and 2012, we included comparable gene transcription data on sea otters from WPWS in 2006, 2007 and 2008 and from the Shumagin Islands on the Alaska Peninsula, Katmai National Park and Preserve, Clam Lagoon on Adak Island in the Aleutian Archipelago, Kodiak Island and captive animals from aquaria (Figure 1).

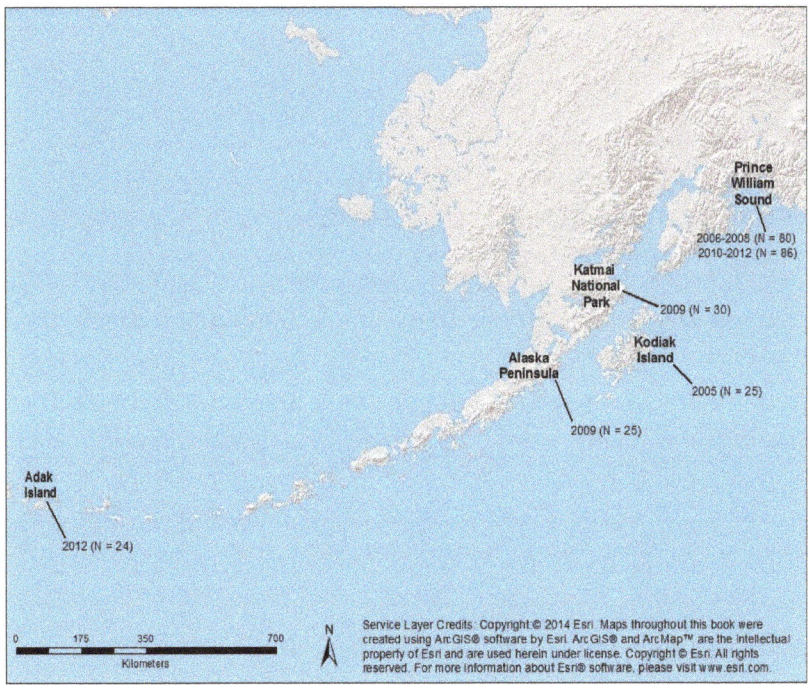

Figure 1. Map of sampling locations in Alaska, including Adak Island (Clam Lagoon, CL), the Alaska Peninsula (AP), Katmai National Park (KAT), Kodiak Island (KOD) and Prince William Sound (PWS1 and PWS2). Captive sampling locations are not shown.

Herein, we provide the results of gene transcription analyses on sea otters sampled at WPWS between 2006 and 2012 and compare these data to those from sea otter populations sampled across southwest Alaska and from aquaria.

2. Materials and Methods

2.1. Sea Otter Samples

Free-ranging sea otters were sampled from five locations: (1) WPWS in 2006, 2007 and 2008 (n = 80) and in 2010 and 2012 (n = 88); (2) the Alaska Peninsula (AP) in 2009 (n = 25); (3) Katmai (KAT) in 2009 (n = 30); (4) Kodiak (KOD) in 2005 (n = 25); and (5) Clam Lagoon (CL) at Adak Island in 2012 (n = 24). Wild sea otters were captured, anesthetized with fentanyl citrate and midazolam hydrochloride [17] and blood drawn by jugular venipuncture within 1–2 hours of the initial capture. Capture methods are presented in detail in Miles *et al.* (2012) [6] and Bodkin *et al.* (2012) [7].

Blood samples from 17 captive reference sea otters were obtained from the Monterey Bay Aquarium (n = 9) (Monterey, CA), Shedd Aquarium (n = 4) (Chicago, IL), Oregon Coast Aquarium (n = 2) (Newport, OR) and the Vancouver Aquarium (n = 2) (Vancouver, BC) in 2008, 2009 and 2010 [5]. These animals were identified as clinically normal by staff veterinarians at these aquaria at the time of blood collection.

2.2. Blood and RNA Processing

A 2.5-mL sample from each sea otter was drawn directly into a PAXgene™ blood RNA collection tube (PreAnalytiX©, Hombrechtikon, Switzerland) from either the jugular or popliteal vein and then frozen at −20 °C until extraction of RNA [5]. The PAXgene tube contains RNA-stabilizing reagents that protect RNA molecules from degradation by RNases and prevent further induction of gene transcription. Without stabilization, copy numbers of individual mRNA species in whole blood can change many-fold during storage and transport. The RNA from blood in PAXgene tubes was isolated according to the manufacturer's standard protocols [18]. All RNA was checked for quality by running on both an agarose gel and on a NanoDrop 2000 (Thermo Fisher Scientific Inc., Wilmington, DE, USA), achieving A260/A280 ratios of approximately 2.0 and A260/A230 ratios of less than 1.0. A standard cDNA synthesis was performed on 1 µg of RNA template from each animal (QuantiTect® Reverse Transcription Kit, Qiagen, Valencia, CA, USA) [18]. Quantitative real-time polymerase chain reaction (qPCR) systems for the individual, sea otter-specific reference or housekeeping gene (S9) and genes of interest (Table 1) were run in separate wells [18]. Amplifications were conducted on a 7300 Real-time Thermal Cycler (Applied Biosystems™, Foster City, CA, USA) with reaction conditions identical to those in Bowen *et al.* (2007, 2012) [5,18] and Miles *et al.* (2012) [6].

Table 1. Documented function of 10 genes identified in free-ranging sea otters sampled at the Alaska Peninsula, Katmai, Kodiak, Clam Lagoon, Prince William Sound 2006–2008, Prince William Sound 1010–2012 and in clinically normal captive sea otters. Amplification efficiencies of all primer pairs were between 90% and 105%.

Gene	Gene Function
HDC	The HDCMB21P gene codes for a translationally-controlled tumor protein (TCTP) implicated in cell growth, cell cycle progression, malignant transformation, tumor progression and in the protection of cells against various stress conditions and apoptosis [19–21]. Environmental triggers may be responsible for population-based up-regulation of HDC. HDC transcription is known to increase with exposure to carcinogenic compounds, such as polycyclic aromatic hydrocarbons [18,22,23].
COX2	Cyclooxygenase-2 catalyzes the production of prostaglandins that are responsible for promoting inflammation [24]. Cox2 is responsible for the conversion of arachidonic acid to prostaglandin H2, a lipoprotein critical to the promotion of inflammation [25]. Upregulation of Cox2 is indicative of cellular or tissue damage and an associated inflammatory response.
CYT	The complement cytolysis inhibitor protects against cell death [26]. Upregulation of CYT is indicative of cell or tissue death.

Table 1. *Cont.*

Gene	Gene Function
AHR	The aryl hydrocarbon receptor responds to classes of environmental toxicants, including polycyclic aromatic hydrocarbons, polyhalogenated hydrocarbons, dibenzofurans and dioxin [27]. Depending on the ligand, AHR signaling can modulate T-regulatory (T_{REG}) (immune-suppressive) or T-helper type 17 (T_H17) (pro-inflammatory) immunologic activity [28,29].
THRβ	The thyroid hormone receptor beta can be used as a mechanistically-based means of characterizing the thyroid-toxic potential of complex contaminant mixtures [30]. Thus, increases in THRβ transcription may indicate exposure to organic compounds, including PCBs, and associated potential health effects, such as developmental abnormalities and neurotoxicity [30,31].
HSP 70	The heat shock protein 70 is produced in response to thermal or other stress, including hyperthermia, oxygen radicals, heavy metals and ethanol [32,33].
IL-18	Interleukin-18 is a pro-inflammatory cytokine [24]. IL-18 plays an important role in inflammation and host defense against microbes [34].
IL-10	Interleukin-10 is an anti-inflammatory cytokine [24]. Levels of IL-10 have been correlated with the relative health of free-ranging harbor porpoises, e.g., increased amounts of IL-10 correlated with chronic disease, whereas the cytokine was relatively reduced in apparently fit animals experiencing acute disease [35]. Association of IL-10 transcription with chronic disease has also been documented in humans [36].
DRB	A component of the major histocompatibility complex, the DRB class II gene is responsible for the binding and presentation of processed antigen to T_H lymphocytes, thereby facilitating the initiation of an immune response [24,37]. Upregulation of MHC genes has been positively correlated with parasite load [37], whereas downregulation of MHC has been associated with contaminant exposure [38,39].
Mx1	The Mx1 gene responds to viral infection [40]. Vertebrates have an early strong innate immune response against viral infection, characterized by the induction and secretion of cytokines that mediate an antiviral state, leading to the upregulation of the MX-1 gene [41].

2.3. Targeted Genes

The 10 genes targeted in our study represent multiple physiological systems that play a role in immuno-modulation, inflammation, cell protection, tumor suppression, cellular stress-response, xenobiotic metabolizing enzymes and antioxidant enzymes. These genes can be modified by biological, physical or anthropogenic impacts and consequently provide information on the general type of stressors present in a given environment (Table 1). Note the inverse relationship in interpretation; *i.e.*, that lower values in Table 2 correspond to higher transcription rates.

2.4. Statistical Analyses

We used nonparametric statistical analyses because the cycle threshold (C_T) measure of gene transcription provided by qPCR may have a lognormal distribution [15]. We used conventional nonparametric mean comparison tests (Kruskal–Wallis with Dunns' multiple comparison; NCSS© Statistical Software, 2007, Kaysville, UT, USA) to evaluate transcript values of each gene by classification groups (7 groups, based on location, including captives as a reference "location" group, and including 2 temporal groups from WPWS). We conducted multivariate, nonparametric, multi-dimensional scaling analysis (NMDS) in conjunction with cluster analysis for statistical and graphical representation of individual sea otters clustered by similarity in transcription and not by pre-defined groups, such as location [42]. Statistical comparisons of individuals grouped by clusters were made using SIMPROF, which is a similarity profile permutation test for significance among *a priori*, unstructured clusters of samples. We used ANOSIM, a nonparametric analogue to a 2-way ANOVA, to test for differences in gene transcription among years, between sexes and among three age groups, *i.e.*, juvenile, adult and aged adult [43]. Statistical significance was based on *p*-values ≤0.05.

3. Results

Gene transcription (C_T) values differed among sea otters sampled in WPWS in 2006, 2007, 2008, 2010 and 2012 (ANOSIM, $p < 0.001$, global $R = 0.594$). When analyzed without *a priori* structure (*i.e.*, year), sea otters separated into two well-defined groups as depicted by NMDS (3D $R = 0.08$; Figure 2) and confirmed by cluster analysis (SIMPROF, $p < 0.001$). These well-defined groups were designated PWS1 (2006, 2007, 2008) and PWS2 (2010, 2012). Transcript patterns were not influenced by sex ($p = 0.08$) or age ($p = 0.16$).

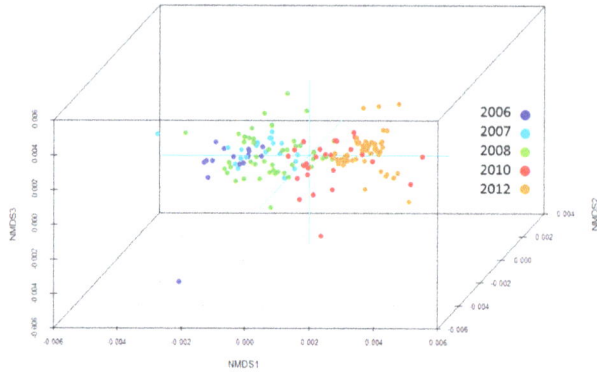

Figure 2. Multivariate, nonparametric, multi-dimensional scaling (NMDS) of gene transcription profiles (see Table 2) of sea otters captured in five different years (2006, 2007, 2008, 2010, 2012) in western Prince William Sound, Alaska, showing distinct separation of 2006–2008 samples from the 2010 and 2012 samples.

For the analysis of all groups, patterns depicted by the NMDS analyses were similar to those reported in Miles *et al.* (2012), with differences attributable to the inclusion of the additional groups (Figure 3). Groups generally separated into three distinctive clusters: (1) KOD and PWS1; (2) CL and PWS2; and (3) KAT, AP and captive sea otters (2D $R = 0.15$; SIMPROF, $p < 0.001$; Figure 3).

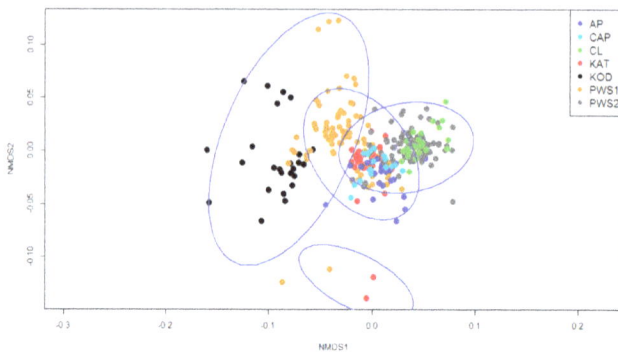

Figure 3. Multivariate, nonparametric, multi-dimensional scaling (NMDS) with cluster analysis of gene transcription profiles (see Table 2) of sea otters sampled at the Alaska Peninsula (AP), Katmai (KAT), Kodiak (KOD), Clam Lagoon (CL), western Prince William Sound 2006, 2007 and 2008 (PWS1), western Prince William Sound 2010 and 2012 (PWS2) and clinically normal captive otters (CAP).

Table 2. Geometric mean (normalized to the S9 housekeeping gene in each animal) cycle threshold (C_T) transcription values (and 95% confidence intervals) for targeted genes (see Table 1) in sea otters sampled at the Alaska Peninsula (AP), Katmai (KAT), Kodiak (KOD), Clam Lagoon (CL), Prince William Sound 2006–2008 (PWS1), Prince William Sound 2010–2012 (PWS2) and clinically normal captive otters (CAP). Letter (a,b,c,d) differences denote significant differences among populations (Kruskal–Wallis with Dunns' multiple comparison); lack of a letter (a,b,c,d) denotes no significant difference from any other group. Note that the smaller the mean value, the higher the level of transcription.

Group	Gene									
	HDC	COX2	CYT	AHR	THRβ	HSP70	IL18	IL10	DRB	MX1
CAP	5.90 [ad] (5.02–6.94)	6.78 [abcd] (6.02–7.64)	2.41 (1.91–3.04)	11.01 [abe] (10.56–11.48)	13.30 [a] (12.49–14.56)	9.62 [ab] (8.74–10.59)	1.65 [acde] (1.05–2.60)	13.70 [a] (13.01–14.44)	−0.33 (−0.86–0.21)	10.99 [ab] (9.95–12.15)
AP	6.26 [a] (5.47–7.16)	6.60 [abcd] (5.92–7.36)	1.90 (1.45–2.52)	10.67 [ab] (9.94–11.46)	13.32 [a] (12.19–14.56)	8.58 [abd] (7.97–9.23)	1.68 [acde] (1.30–2.17)	13.21 [a] (12.21–14.30)	−0.91 [a] (−1.57–−0.26)	12.61 [a] (11.42–13.93)
KAT	4.54 [ab] (4.06–5.08)	7.68 [bd] (7.10–8.30)	1.96 (1.54–2.50)	10.36 [ab] (9.79–10.96)	12.53 [a] (11.86–13.23)	8.26 [abd] (7.56–9.04)	2.78 [ce] (2.15–3.59)	13.45 [a] (12.62–14.33)	−0.56 [a] (−1.09–−0.03)	12.56 [a] (11.99–13.16)
KOD	−1.84 [e] (−2.33–−1.35)	5.44 [c] (4.79–6.16)	2.59 (2.04–3.28)	8.80 [d] (8.09–9.57)	9.50 [c] (8.62–10.46)	5.48 [d] (4.86–6.18)	5.19 [b] (4.61–5.85)	9.03 [c] (8.26–9.87)	1.29 (0.78–2.13)	8.26 [b] (7.64–8.92)
CL	10.30 [c] (10.06–10.54)	9.45 [e] (8.91–10.02)	1.53 (1.28–1.83)	12.78 [c] (12.38–13.18)	16.85 [b] (15.83–17.93)	14.07 [c] (13.17–15.02)	2.49 [acde] (2.06–3.01)	22.09 [b] (20.85–23.40)	0.43 [b] (0.22–0.84)	16.89 [c] (15.36–18.57)
PWS1	4.01 [bde] (2.89–5.57)	7.98 [ad] (7.60–8.38)	1.88 (1.60–2.21)	10.17 [abd] (9.75–10.62)	11.35 [ac] (10.74–11.99)	9.76 [ab] (9.17–10.39)	1.60 [d] (1.24–2.07)	13.34 [a] (12.77–13.94)	1.08 [b] (0.80–1.44)	10.41 [b] (10.05–10.78)
PWS2	8.94 [c] (8.46–9.46)	9.30 [e] (8.96–9.65)	1.62 (1.45–1.80)	12.07 [c] (11.79–12.36)	16.09 [b] (15.54–16.66)	13.62 [c] (13.10–14.01)	2.38 [e] (2.21–2.56)	20.28 [b] (19.44–21.16)	−0.071 (−0.25–0.10)	14.95 [c] (14.50–15.41)

Overall, gene transcription (C_T) values differed among groups (Figure 3). The transcript profiles from the AP, KAT and clinically normal captive groups were similar and differed from the other three groups. Profiles of the PWS2 and CL groups were similar. In general, gene transcription patterns in the PWS1 group of sea otters (captured 2006–2008) were indicative of molecular reactions to organic exposure, tumor formation, inflammation and viral infection that may be consistent with chronic, low-grade exposure to an organic substance (Tables 1 and 2). Although the KOD group overlapped with PWS1 in the NMDS analysis (Figure 3), the transcription of seven genes was highly upregulated (at least >2 C_T values rounded) at KOD compared to PWS1 (Table 2). The PWS2 group (captured 2010 and 2012), in contrast, had a general pattern of lower transcription, with eight of the 10 genes showing significant downregulation compared to PWS1. The PWS2 sea otters grouped statistically with the CL sea otters (Figure 3).

Using Kruskal–Wallis, nine of the ten genes evaluated had significant differences between at least two classification groups; only CYT did not differ among groups (Table 2). Geometric mean transcript values were highest (*i.e.*, lowest C_T values) at KOD for seven of the nine genes showing significant differences (HDC, COX2, AHR, THRβ, HSP70, IL10, MX1). Geometric mean transcript values for IL18 were highest in the PWS1, AP and CAP groups. The lowest geometric mean transcript values among groups generally were found in CL and PWS2 sea otters for seven of the nine genes (HDC, COX2, AHR, IL10, MX1 at CL and THRβ, HSP70 at PWS2). The lowest geometric mean transcript value for IL18 was in the KOD group. The largest ranges of geometric means among groups (most variable expression) were identified for HDC and IL10, while the smallest ranges occurred for DRB, IL18 and CYT (with CYT showing no significant variation among any groups). Genes with larger ranges may be subject to greater environmental variation in a particular system than genes with smaller ranges.

4. Discussion

The genes analyzed in our study can be grouped into functional categories that include immuno-modulation, pathogen response, inflammation, cell signaling, xenobiotic metabolizing enzymes and cellular stress response (see Table 1). Although transcription studies generally focus on genes that are differentially transcribed among groups, genes that show no difference among groups are also of importance. Of particular note in this study was the lack of statistical difference in gene transcription between the AP and clinically normal captive sea otters (see Table 2 and Figure 3).

The interpretation of the high similarity of wild-captured sea otters to documented clinically normal, healthy sea otters is that individuals in the AP subpopulation are healthy and are not subject to substantial hydrocarbon exposure, disease or food limitation. Transcript patterns from the KAT subpopulation of sea otters also were similar to those of the AP and captive populations (Figure 3). These interpretations are further supported by population status and trajectory data, indicating that both the KAT and AP populations of sea otters are below carrying capacity [44,45].

Two other groups with remarkably similar transcript patterns were CL and PWS2, both exhibiting relatively low levels of transcription in most genes examined. Relatively low levels of select gene transcripts have been described in mice experiencing a nutritional deficit [46]. Alternatively, low transcription may be the result of unbalanced physiological resource allocation. For example, immune defenses exist to impede infections, but other ecological demands (e.g., stressors related to nutrition, weather and predation) can supersede this, causing immune defenses to be compromised [47]. This interpretation is consistent with data on rates of energy recovery of various sea otter populations, indicating that food resources for sea otters at CL and in WPWS (2010, 2012) were limited, compared to other reference groups sampled in this study ([46,48]). The population status of stable or near carrying capacity for both CL and PWS2 [46,49] further supports the potential of limited nutritional resources in these groups.

Distinct transcript patterns also existed among groups and reflect the influence of environmental factors, potentially including food availability, contaminants, disease and predation. Within a group, there will also be behavioral differences (e.g., foraging patterns, home range) among sea otters that may contribute to variation in gene transcripts. For example, Bodkin *et al.* (2012) demonstrated marked differences among sea otters in the extent of intertidal foraging and, thus, potential exposure to lingering oil. Consequently, we expect that some sea otters in WPWS may have minimal exposure to lingering oil and transcript patterns that appear clinically normal. This is supported by some amount of overlap in transcript profiles, as noted in Figure 3.

Interestingly, transcriptional differences of sea otters from KOD and PWS1 compared to the other groups were evident, and transcription levels in sea otters from KOD, in particular, were high in relation to those of other groups. The PWS1 and KOD groups appeared to have immunological or physiological responses that indicated greater organic compound exposure relative to the other populations examined, but their profile motifs differed, suggesting unique environmental stressors at each site. Genomic profiling has successfully linked specific signatures to unique combinations of chemical contaminants in other species [50–53]. In fact, the transcription profile of the KOD otters is more consistent with that of a dioxin-induced profile, while the transcription profile of PWS1 otters (in particular, those from the area that received the heaviest shoreline oiling in 1989) is more consistent with a polycyclic aromatic hydrocarbon (PAH)-induced profile [27]. It is noteworthy that AHR transcription was highest at KOD, followed by PWS1, although the latter did not differ significantly from the other groups. Upregulation of AHR is indicative of current exposure to classes of environmental toxicants, including PAHs, polyhalogenated hydrocarbons, dibenzofurans and dioxin [28]. Chronic exposure to specific toxicants may not necessarily cause a sustained increase in AHR transcription [6,15], but can be associated with potential downstream consequences (e.g., modulation of T-regulatory (T_{REG}) (immune-suppressive) or T-helper type 17 ($T_{H}17$) (pro-inflammatory) immunologic activity [29,39]); however, T-regulatory cell activity was not specifically analyzed in this study. The transcript profile in PWS1 sea otters appears consistent with findings from Bodkin *et al.* (2012) [7] indicating that foraging sea otters in WPWS during that time period were subject to ongoing, potentially intermittent, exposure to lingering oil in the environment. Further, sea otters from the spill area in WPWS in 2008 demonstrated elevated transcription of several genes, including HDC and THRβ, and downregulation of the DRB gene; a similar pattern for these three genes was seen at KOD. Dong *et al.* (1997) [54] reported downregulation of DRB by a dioxin compound, and both polycyclic aromatic hydrocarbons (constituents of crude oil) and dioxin-like compounds have been implicated in physiologic detoxification responses.

In summary, gene transcript profiles suggest that in 2008, sea otters in WPWS were still subject to lingering oil exposure. This finding was consistent with studies that quantified oil encounter rates by foraging sea otters in WPWS ranging from 2–24 times per year and documented the presence of PAHs in sea otter forage pits prior to 2009 [7]. Interpretation of the 2012 gene transcription profiles of WPWS sea otters is complicated by general low levels of transcription. The low transcript levels seen in WPWS (2010, 2012) and in CL sea otters could be consistent with an inability to mount effective responses to pathogens, contaminants, injury or other stressors when compared to earlier time intervals or other groups. In effect, the overall dampening of the molecular response precludes determination of whether or not WPWS sea otters showed a continued response to lingering oil in 2010–2012.

However, several studies on sea otter demographics indicated that by 2012, the WPWS sea otter population had returned to pre-spill conditions. While sea otter abundance at the scale of WPWS had demonstrated modest increases since 1993, areas most severely impacted by oil-related mortality did not return to pre-spill numbers until 2011 [8]. The numerical recovery of sea otters was supported by improved survival of sea otters after 2009, with a return to rates observed prior to the spill [4,8,55]. The findings for sea otters related to diminished oil exposure and population recovery were consistent with related findings for sea ducks. Prior to 2009, data indicated continued exposure to two species of nearshore sea ducks, with diminished exposure to oil evident in Barrows goldeneye by 2010 [3] and Harlequin ducks by 2013 [56]. An expanded study of the broader sea otter transcriptome would further the identification of environmental stressors responsible for the overall low levels of gene transcripts observed in WPWS in 2010 and 2012.

Acknowledgments: This research was funded in part by the *Exxon Valdez* Oil Spill Trustee Council; however, the findings and conclusions do not necessarily reflect the views or position of the Trustee Council. Funding was also provided by the U.S. Geological Survey (USGS) Western Ecological Research Center and the USGS Alaska Science Center. We gratefully acknowledge the assistance of Heather Coletti, George Esslinger, KimKloecker, Daniel Monson, Ben Weitzman and Joe Tomoleoni, USGS, and Michael Murray and Marissa Viens, Monterey Bay Aquarium, for sample collections. We thank Robin Keister for laboratory assistance. Samples were collected under permits authorized by the U.S. Fish and Wildlife Service and the approval of the Animal Care and Use Committee of the USGS Alaska Science Center. Any use of trade, firm or product names is for descriptive purposes only and does not imply endorsement by the U.S. Government.

Author Contributions: All authors contributed equally to the project.

Conflicts of Interest: The authors declare no conflict of interest.

References

1. Bodkin, J.L.; Esler, D.; Rice, S.D.; Matkin, C.O.; Ballachey, B.E. The effects of spilled oil on coastal ecosystems: Lessons from the Exxon Valdez spill. In *Coastal Conservation*; Maslo, B., Lockwood, J.L., Eds.; Cambridge University Press: New York, NY, USA, 2014; pp. 311–346.

2. Esler, D.; Trust, K.A.; Ballachey, B.E.; Iverson, S.A.; Lewis, T.L.; Rizzolo, D.J.; Mulcahy, D.M.; Miles, A.K.; Woodin, B.R.; Stegeman, J.J.; *et al.* Cytochrome P4501 A biomarker indication of oil exposure in harlequin ducks up to 20 years after the Exxon Valdez oil spill. *Environ. Toxicol. Chem.* **2010**, *29*, 1138–1145. [PubMed]

3. Esler, D.; Ballachey, B.E.; Trust, K.A.; Iverson, S.A.; Reed, J.A.; Miles, A.K.; Henderson, J.D.; Woodin, B.R.; Statesman, J.J.; McAdie, M.; *et al.* Cytochrome P4501 A biomarker indication of the timeline of chronic exposure of Barrow's goldeneyes to residual Exxon Valdez oil. *Mar. Pollut. Bull.* **2011**, *62*, 609–614. [CrossRef] [PubMed]

4. Monson, D.H.; Doak, D.F.; Ballachey, B.E.; Bodkin, J.L. Could residual oil from the *Exxon Valdez* spill create a long-term population 'sink' for sea otters in Alaska? *Ecol. Appl.* **2011**, *21*, 2917–2932. [CrossRef]

5. Bowen, L.; Miles, A.K.; Murray, M.; Haulena, M.; Tuttle, J.; Van Bonn, W.; Adams, L.; Bodkin, J.L.; Ballachey, B.E.; Estes, J.A.; *et al.* Gene transcription in sea otters (*Enhydra. lutris*); emerging diagnostics in marine mammal and ecosystem health. *Mol. Ecol. Resour.* **2012**, *12*, 67–74. [CrossRef] [PubMed]

6. Miles, A.K.; Bowen, L.; Ballachey, B.E.; Bodkin, J.L.; Murray, M.; Estes, J.L.; Keister, R.A.; Stott, J.L. Variation in transcript profiles in sea otters (*Enhydra lutris*) from Prince William Sound, Alaska and clinically normal reference otters. *Mar. Ecol. Prog. Ser.* **2012**, *451*, 201–212. [CrossRef]

7. Bodkin, J.L.; Ballachey, B.E.; Coletti, H.A.; Esslinger, G.G.; Kloecker, K.A.; Rice, S.D.; Reed, J.A.; Monson, D.H. Long-term effects of the *Exxon Valdez* oil spill—Sea otter foraging in the intertidal as a pathway of exposure to lingering oil. *Mar. Ecol. Prog. Ser.* **2012**, *447*, 273–287. [CrossRef]

8. Ballachey, B.E.; Monson, D.H.; Esslinger, G.G.; Kloecker, K.; Bodkin, J.; Bowen, L.; Miles, A.K. 2013 Update on Sea Otter Studies to Assess Recovery from the 1989 Exxon Valdez Oil Spill, Prince William Sound, Alaska: U.S. Geological Survey Open-File Report 2014–1030, 40p. Available online: http://dx.doi.org/10.3133/ofr20141030 (accessed on 20 May 2016).

9. Ballachey, B.E.; Bodkin, J.L.; Esler, D.; Rice, S.D. Lessons from the 1989 *Exxon Valdez* oil spill: A biological perspective. In *Impacts of Oil Spill Disasters on Marine Habitats and Fisheries in North America*; Alford, J.B., Peterson, M.S., Green, C.C., Eds.; CRC Press: Boca Raton, FL, USA, 2014; pp. 181–198.

10. Harwell, M.A.; Gentile, J.H.; Johnson, C.B.; Garshelis, D.L.; Parker, K.R. A Quantitative Ecological Risk Assessment of the Toxicological Risks from Exxon Valdez Subsurface Oil Residues to Sea Otters at Northern Knight Island, Prince William Sound, Alaska. *Hum. Ecol. Risk Assess.* **2010**, *16*, 727–761. [CrossRef] [PubMed]

11. Ballachey, B.E.; Bodkin, J.L.; Monson, D.H. Quantifying long-term risks to sea otters from the 1989 'Exxon Valdez' oil spill: Reply to Harwell and Gentile (2013). *Mar. Ecol. Prog. Ser.* **2013**, *480*, 297–301. [CrossRef]

12. Harwell, M.A.; Gentile, J.H. Assessing risks to sea otters and the Exxon Valdez oil spill: New scenarios, attributable risk, and recovery. *Hum. Ecol. Risk Assess.* **2014**, *20*, 889–916. [CrossRef] [PubMed]

13. Peterson, C.H.; Rice, S.D.; Short, J.W.; Esler, D.; Bodkin, J.L.; Ballachey, B.E.; Irons, D.B. Long-term ecosystem response to the Exxon Valdez oil spill. *Science* **2003**, *302*, 2082–2086. [CrossRef] [PubMed]

14. Farr, S.; Dunn, R.T. Concise review: Gene expression applied to toxicology. *Toxicol. Sci.* **1999**, *50*, 1–9. [CrossRef] [PubMed]

15. McLoughlin, K.; Turteltaub, K.; Bankaitis-Davis, D.; Gerren, R.; Siconolfi, L.; Storm, K.; Cheronis, J.; Trollinger, D.; Macejak, D.; Tryon, V.; *et al.* Limited dynamic range of immune response gene expression observed in healthy blood donors using RT-PCR. *J. Mol. Med.* **2006**, *12*, 185–195.

16. Poynton, H.C.; Vulpe, C.D. Ecotoxicogenomics: Emerging technologies for emerging contaminants. *J. Am. Water Resour. Assoc.* **2009**, *45*, 83–96. [CrossRef]

17. Monson, D.H.; McCormick, C.; Ballachey, B.E. Chemical anesthesia of northern sea otters (*Enhydra lutris*)—Results of past field studies. *J. Zoo Wildl. Med.* **2001**, *32*, 181–189. [PubMed]

18. Bowen, L.; Schwartz, J.; Aldridge, B.; Riva, F.; Miles, A.K.; Mohr, F.C.; Stott, J.L. Differential gene expression induced by exposure of captive mink to fuel oil—A model for the sea otter. *EcoHealth* **2007**, *4*, 298–309. [CrossRef]

19. Bommer, U.A.; Thiele, B.J. The translationally controlled tumour protein (TCTP). *Int. J. Biochem. Cell Biol.* **2004**, *36*, 379–385. [CrossRef]

20. Tuynder, M.; Fiucci, G.; Prieur, S.; Lespagnol, A.; Geant, A.; Beaucourt, S.; Duflaut, D.; Besse, S.; Susini, L.; Cavarelli, J.; *et al.* Translationally controlled tumor protein is a target of tumor reversion. *Proc. Natl. Acad. Sci. USA* **2004**, *101*, 15364–15369. [CrossRef] [PubMed]

21. Ma, Q.; Geng, Y.; Xu, W.; Wu, Y.; He, F.; Shu, W.; Huang, M.; Du, H.; Li, M. The role of translationally controlled tumor protein in tumor growth and metastasis of colon adenocarcinoma cells. *J. Proteome Res.* **2010**, *9*, 40–49. [CrossRef] [PubMed]

22. Raisuddin, S.; Kwok, K.W.H.; Leung, K.M.Y.; Schlenk, D.; Lee, J. The copepod *Tigriopus*—A promising marine model organism for ecotoxicology and environmental genomics. *Aquat. Toxicol.* **2007**, *83*, 161–173. [CrossRef] [PubMed]

23. Zheng, S.; Song, Y.; Qiu, X.; Sun, T.; Ackland, M.L.; Zhang, W. Annetocin and TCTP expressions in the earthworm *Eisenia fetida* exposed to PAHs in artificial soil. *Ecotoxicol. Environ. Saf.* **2008**, *71*, 566–573. [CrossRef] [PubMed]

24. Goldsby, R.A.; Kindt, T.J.; Osborne, B.A.; Kuby, J. *Immunology*, 5th ed.; W.H. Freeman and Company: New York, NY, USA, 2003.

25. Harris, S.G.; Padilla, J.; Koumas, L.; Ray, D.; Phipps, R.P. Prostaglandins as modulators of immunity. *Trends Immunol.* **2002**, *23*, 144–150. [CrossRef]

26. Jenne, D.E.; Tschopp, J. Molecular structure and functional characterization of a human complement cytolysis inhibitor found in blood and seminal plasma—Identity to sulfated glycoprotein 2, a constituent of rat testis fluid. *Proc. Natl. Acad. Sci. USA* **1989**, *86*, 7123–7127. [CrossRef] [PubMed]

27. Zeytun, A.; McKallip, R.J.; Fisher, M.; Camacho, I.; Nagarkatti, M.; Nagarkatti, P.S. Analysis of 2,3,7,8-tetrachlorodibenzo-p-dioxin-induced gene expression profile *in vivo* using pathway-specific cDNA arrays. *Toxicology* **2002**, *23*, 241–260. [CrossRef]

28. Oesch-Bartlomowicz, B.; Huelster, A.; Wiss, O.; Antoniou-Lipfert, P.; Dietrich, C.; Arand, M.; Weiss, C.; Bockamp, E.; Oesch, F. Aryl hydrocarbon receptor activation by cAMP *vs.* dioxin: Divergent signaling pathways. *Proc. Natl. Acad. Sci. USA* **2005**, *102*, 9218–9223. [CrossRef] [PubMed]

29. Quintana, F.J.; Basso, A.S.; Iglesias, A.H.; Korn, T.; Farez, M.F.; Bettelli, E.; Caccamo, M.; Oukka, M.; Weiner, H.L. Control of T(reg) and T(H)17 cell differentiation by the aryl hydrocarbon receptor. *Nature* **2008**, *453*, 6–7. [CrossRef] [PubMed]

30. Tabuchi, M.; Veldhoen, N.; Dangerfield, N.; Jeffries, S.; Helbing, C.C.; Ross, P.S. PCB-related alteration of thyroid hormones and thyroid hormone receptor gene expression in free-ranging harbor seals (*Phoca vitulina*). *Environ. Health Perspect.* **2006**, *114*, 1024–1031. [CrossRef] [PubMed]

31. Tsai, M.J.; O'Malley, B.W. Molecular mechanisms of action of steroid/thyroid receptor superfamily members. *Annu. Rev. Biochem.* **1994**, *63*, 451–486. [CrossRef] [PubMed]

32. Iwama, G.K.; Mathilakath, M.V.; Forsyth, R.B.; Ackerman, P.A. Heat shock proteins and physiological stress in fish. *Am. Zool.* **1999**, *39*, 901–909. [CrossRef]

33. Tsan, M.; Gao, B. Cytokine function of heat shock proteins. *Am. J. Physiol. Cell Physiol.* **2004**, *286*, 739–744. [CrossRef] [PubMed]

34. Krumm, B.; Meng, X.; Li, Y.; Xiang, Y.; Deng, J. Structural basis for antagonism of Human interleukin 18 by poxvirus interleukin 18-binding protein. *Proc. Natl. Acad. Sci. USA* **2008**, *105*, 20711–20715. [CrossRef] [PubMed]

35. Beineke, A.; Siebert, U.; Muller, G.; Baumgartner, W. Increased blood interleukin-10 mRNA levels in diseased free-ranging harbor porpoises (*Phocoena phocoena*). *Vet. Immunol. Immunopathol.* **2007**, *115*, 100–106. [CrossRef] [PubMed]

36. Rigopoulou, E.I.; Abbott, W.G.; Haigh, P.; Naoumov, N.V. Blocking of interleukin-10 receptor-a novel approach to stimulate T-helper cell type 1 responses to hepatitis C virus. *Clin. Immunol.* **2005**, *117*, 57–64. [CrossRef] [PubMed]

37. Bowen, L.; Aldridge, B.; Miles, A.K.; Stott, J.L. Expressed MHC class II genes in sea otters (*Enhydra lutris*) from geographically disparate populations. *Tissue Antigens* **2006**, *67*, 402–408. [CrossRef] [PubMed]

38. Wegner, K.M.; Kalbe, M.; Rauch, G.; Kurtz, J.; Schaschl, H.; Reusch, T.B.H. Genetic variation in MHC class II expression and interactions with MHC sequence polymorphism in three-spined sticklebacks. *Mol. Ecol.* **2006**, *15*, 1153–1164. [CrossRef] [PubMed]

39. Veldhoen, M.; Hirota, K.; Westendorf, A.M.; Buer, J.; Dumoutier, L.; Renauld, J.C.; Stockinger, B. The aryl hydrocarbon receptor links TH17-cell-mediated autoimmunity to environmental toxins. *Nature* **2008**, *453*, 106–109. [CrossRef] [PubMed]

40. Tumpey, T.M.; Szretter, K.J.; Van Hoeven, N.; Katz, J.M.; Kochs, G.; Haller, O.; Garcia-Sister, A.; Staeheli, P. The Mx1 gene protects mice against the pandemic 1918 and highly lethal human H5 N1 influenza viruses. *J. Virol.* **2007**, *81*, 10818–10821. [CrossRef] [PubMed]

41. Kibenge, M.J.T.; Munir, K.; Kibenge, F.S.B. Constitutive expression of Atlantic salmon Mx1 protein in CHSE-214 cells confers resistance to infectious salmon Anaemia virus. *J. Virol.* **2005**, *2*, 75. [CrossRef] [PubMed]

42. R Development Core Team. *R: A Language and Environment for Statistical Computing*; ISBN: 3-900051-07-0. R Foundation for Statistical Computing: Vienna, Austria, 2012; Available online: http://www.R-project.org/ (accessed on 20 May 2016).

43. Monson, D.H.; Doak, B.E.; Johnson, A.; Bodkin, J.L. Long-term impact of the Exxon Valdez oil spill on sea otters, assessed through age-dependent mortality patterns. *Proc. Natl. Acad. Sci. USA* **2000**, *97*, 6562–6567. [CrossRef] [PubMed]

44. U.S. Fish and Wildlife Service. *Northern Sea Otter (Enhydra lutris kenyoni) Southwest Alaska Stock Report*; U.S. Fish and Wildlife Service: Anchorage, AK, USA, 2014.

45. Coletti, H.A.; Bodkin, J.L.; Monson, D.H.; Dean, T.A.; Ballachey, B.E. Engaging form and function to detect and infer cause of change in an Alaska marine ecosystem. *Ecosphere* **2016**. submitted.

46. Saucillo, D.C.; Gerriets, V.A.; Sheng, J.; Rathmell, J.C.; MacIver, N.J. Leptin metabolically licenses T cells for activation to link nutrition and immunity. *J. Immunol.* **2014**, *192*, 136–144. [CrossRef] [PubMed]

47. Martin, L.B.; Hopkins, W.A.; Mydlarz, L.D.; Rohr, J.R. The effects of anthropogenic global changes on immune functions and disease resistance. *Ann. N. Y. Acad. Sci.* **2010**, *1195*, 129–148. [CrossRef] [PubMed]

48. Bodkin, J.L.; U.S. Geological Survey, Port Townsend, WA, USA. Unpublished work. 2016.

49. Estes, J.A.; Tinker, M.T.; Williams, T.M.; Doak, D.F. Killer whale predation on sea otters linking oceanic and nearshore ecosystems. *Science* **1998**, *282*, 473–476. [CrossRef] [PubMed]

50. Menzel, R.; Swain, S.C.; Hoess, S.; Claus, E.; Menzel, S.; Steinberg, C.E.W.; Reifferscheid, G.; Sturzenbaum, S.R. Gene expression profiling to characterize sediment toxicity—A pilot study using Caenorhabditis elegans whole genome microarrays. *BMC Genomics* **2009**, *10*, 160. [CrossRef] [PubMed]

51. Steinberg, C.E.; Sturzenbaum, S.R.; Menzel, R. Genes and environment—Striking the fine balance between sophisticated biomonitoring and true functional environmental genomics. *Sci. Total Environ.* **2008**, *400*, 142–161. [CrossRef] [PubMed]

52. Yang, L.; Kemadjou, J.R.; Zinsmeister, C.; Bauer, M.; Legradi, J.; Muller, F.; Pankratz, M.; Jakel, J.; Strahle, U. Transcriptional profiling reveals barcode-like toxicogenomic responses in the zebrafish embryo. *Genome Biol.* **2007**, *8*, R227. [CrossRef] [PubMed]

53. Poynton, H.C.; Zuzow, R.; Loguinov, A.V.; Perkins, E.J.; Vulpe, C.D. Gene expression profiling in Daphnia magna, Part II: Validation of a copper specific gene expression signature with effluent from two copper mines in California. *Environ. Sci. Technol.* **2008**, *42*, 6257–6263. [CrossRef] [PubMed]

54. Dong, L.; Ma, Q.; Whitlock, J.P., Jr. Down-regulation of major histocompatibility complex Q1 b gene expression by 2,3,7,8-tetrachlorodibenzo-p-dioxin. *J. Biol. Chem.* **1997**, *272*, 29614–29619. [CrossRef] [PubMed]

55. Ballachey, B.E.; Bodkin, J.L. Challenges to Sea Otter Recovery and Conservation. In *Sea Otter Conservation*; Larson, S.E., Bodkin, J.L., VanBlaricom, G.R., Eds.; Elsevier: London, UK, 2015; pp. 63–88.

56. Bowen, L.; Miles, A.K.; Ballachey, B.E.; Esler, D. Long-term Monitoring Program—Evaluating Chronic Exposure of Harlequin Ducks and Sea Otters to Lingering *Exxon Valdez* Oil in Western Prince William Sound. In *Exxon Valdez Oil Spill Trustee Council Restoration Project Final Report (Project 14120114-Q)*; U.S. Geological Survey, Alaska Science Center: Anchorage, AK, USA, 2014.

MDPI AG

St. Alban-Anlage 66

4052 Basel, Switzerland

Tel. +41 61 683 77 34

Fax +41 61 302 89 18

http://www.mdpi.com

JMSE Editorial Office

E-mail: jmse@mdpi.com

http://www.mdpi.com/journal/jmse

www.ingramcontent.com/pod-product-compliance
Lightning Source LLC
Chambersburg PA
CBHW051859210326
41597CB00033B/5954